African Seed Enterprises

Sowing the Seeds of Food Security

African Seed Enterprises

Sowing the Seeds of Food Security

Edited by

Paul Van Mele
Africa Rice Center (AfricaRice)

Jeffery W. Bentley
Bolivia

Robert G. Guéi
Food and Agriculture Organization of the United Nations
Italy

Published by
the Food and Agriculture Organization of the United Nations

and

the Africa Rice Center (AfricaRice)
by arrangement with
CAB International

www.cabi.org

CABI is a trading name of CAB International

CABI Head Office
Nosworthy Way
Wallingford
Oxfordshire OX10 8DE
UK

Tel: +44 (0)1491 832111
Fax: +44 (0)1491 833508
E-mail: cabi@cabi.org
Website: www.cabi.org

CABI North American Office
875 Massachusetts Avenue
7th Floor
Cambridge, MA 02139
USA

Tel: +1 617 395 4056
Fax: +1 617 354 6875
E-mail: cabi-nao@cabi.org

A catalogue record for this book is available from the British Library, London, UK.

Library of Congress Cataloging-in-Publication Data

African seed enterprises : sowing the seeds of food security / edited by Paul Van Mele, Jeffery W. Bentley, Robert G. Guéi.
p. cm.
Includes bibliographical references and index.
ISBN 978-1-84593-843-7 (alk. paper) -- ISBN 978-1-84593-863-5 (pbk. : alk. paper) -- ISBN 978-9-25106-687-4 (fao pbk. : alk. paper) 1. Seed industry and trade--Africa--Case studies. 2. Seeds--Economic aspects--Africa. 3. Seeds--Africa--Marketing. 4. Food crops--Seeds--Africa. I. Mele, Paul van. II. Bentley, Jeffery W. (Jeffery Westwood), 1955- III. Guéi, Robert G. IV. Title.

HD9019.S432A353 2011
338.1'7--dc22

2011003331

Cover design: Marcella Vrolijks
Map production and design: Chabi Adeyemi and Marcella Vrolijks

ISBN-13: 978 1 84593 843 7 (CABI hardback)
ISBN-13: 978 1 84593 863 5 (Africa Rice paperback)
ISBN-13: 978 9 25106 687 4 (FAO paperback)

Correct citation: Van Mele, P., Bentley, J.W. and Guéi, R.G. (eds) 2011. African Seed Enterprises: Sowing the Seeds of Food Security. CAB International, Wallingford, UK, 256 pp.

Commissioning editor: Nigel Farrar
Production editor: Fiona Chippendale

Typeset by SPi, Pondicherry, India.
Printed and bound by Gutenberg Press, Malta.

FSC www.fsc.org MIX Paper from responsible sources FSC® C022612

The paper used for this book is FSC-certified and totally chlorine-free. FSC (the Forest Stewardship Council) is an international network to promote responsible management of the world's forests.

Contents

Contributors

The Editors

Paul Van Mele is an agricultural scientist from Ghent University, Belgium, who obtained his PhD studying Vietnamese fruit farmers' knowledge of pests, natural enemies and pesticides (Wageningen University, 2000). He wrote *Ants as Friends* (2003) and edited *Way out of the Woods* (2003) and *Innovations in Rural Extension* (2005). In 2004, his video project with rural women in Bangladesh won an international award for effective communication. From 2005 to 2010, Paul Van Mele lived in Benin, working as programme leader, Learning and Innovation Systems at AfricaRice. His work with farmers, video and rural radio earned him the 2009 CGIAR Science Award for outstanding communications. He currently runs his own enterprise Agro-Insight, based in Belgium.

Jeffery W. Bentley is an agricultural anthropologist (PhD, University of Arizona, 1986). He lived in Portugal for a year, doing participant observation with smallholder farmers (1983–84). He spent 7 years in Honduras (1987–94) at Zamorano, an agricultural college, helping researchers and smallholders create appropriate integrated pest management technology. He is the author of various articles, books and manuals, including *Today There Is No Misery* (1992) and *Manual for Collaborative Research with Smallholder Farmers* (with Peter Baker, 2002). Jeff Bentley has lived in Bolivia since 1994 and is an international consultant.

Robert G. Guéi is a plant breeder and a geneticist (PhD, Kansas State University, 1991). He spent 14 years at AfricaRice (formerly called WARDA), initially as rice breeder, then as coordinator of the international network of genetic evaluation of rice and finally as Head of the Genetic Resources Unit. Since 2005, he has worked at FAO as an agricultural officer and seed expert. He lives in Rome and has a global responsibility to help countries develop their seed systems. He is the author of various scientific papers, book chapters and manuals, and technical and policy guides, including several on community seed enterprise development. He recently edited the *NERICA Rice Compendium*.

Authors

Adelugba, Kehinde, *National Agricultural Seed Council (NASC), Abuja, Nigeria; E-mail: kaadelugba@yahoo.com*

Ajayi, Olupomi, *Africa Rice Center (AfricaRice), Ibadan, Nigeria; E-mail: o.j.ajayi-ng@cgiar.org*

Almekinders, Conny, *Wageningen University and Research Centre (WUR), Wageningen, The Netherlands; E-mail: conny.almekinders@wur.nl*

Anyang, Robert, *Africa Rice Center (AfricaRice), Dar es Salaam, Tanzania; E-mail: ranyang@hotmail.com*

Bah, Saidu, *Africa Rice Center (AfricaRice), Cotonou, Benin; E-mail: s.bah@cgiar.org*

Barker, Ian, *International Potato Center (CIP), Nairobi, Kenya; E-mail: i.barker@cgiar.org*

Barra, Abdoulaye, *Ministry of Agriculture and Rural Development, Garoua, Cameroon; E-mail: barra_2@yahoo.fr*

Barry, M. Billo, *Institut de recherche agronomique de Guinée (IRAG), Conakry, Guinea; E-mail: billobarry@hotmail.com*

Bentley, Jeffery W., *Independent consultant, Cochabamba, Bolivia; E-mail: jefferywbentley@hotmail.com*

Borus, Dinah, *International Potato Center (CIP), Nairobi, Kenya; E-mail: d.borus@cgiar.org*

Dalohoun, Daniel N., *Africa Rice Center (AfricaRice), Cotonou, Benin; E-mail: d.dalahoun@cgiar.org*

Diallo, Dioukamady, *Laboratoire des Semences (LABOSEM), Sotuba, Mali; E-mail: tassydiallo@yahoo.fr*

Diawara, Sékou, *Institut de recherche agronomique de Guinée (IRAG), Conakry, Guinea; E-mail: sekoudiawara@yahoo.fr*

Drammeh, Essa, *National Agricultural Research Institute (NARI), Sapu, The Gambia; E-mail: essa_drammeh@yahoo.com*

Guéi, Robert G., *Food and Agriculture Organization (FAO), Rome, Italy; E-mail: Gouantoueu.Guei@fao.org*

Guindo, Hamidou, *Direction Régional Agriculture, Gao, Mali; E-mail: hamidguindo@yahoo.fr*

Lyamani, Abderrahmane, *Kénitra, Morocco, E-mail: ablyamani@gmail.com*

Misiko, Michael, *Africa Rice Center (AfricaRice), Cotonou, Benin; E-mail: m.misiko@cgiar.org*

Mukalama, John, *International Center for Tropical Agriculture (CIAT), Maseno, Kenya; E-mail: j.mukalama@cgiar.org*

Oggema, Judith, *Agricultural Development Corporation (ADC)*, Molo, Kenya; *E-mail: judynata@gmail.com*

Okry, Florent, *Africa Rice Center (AfricaRice), Cotonou, Benin; E-mail: okryflorent@yahoo.fr*

Rabeson, Raymond, *FOFIFA, Antananarivo, Madagascar; E-mail: rabeson_raymond@moov.mg*

Ralisoa, Noroseheno, *FIFAMANOR, Andranomanelatra Antsirabe, Madagascar; E-mail: ralisoo@yahoo.fr*

Randriamilandy, Ketamalala, *Service Officiel de Contrôle des Semences (SOC), Antananarivo, Madagascar; E-mail: ketamalal@yahoo.fr*

Reece, J. David, *Africa Rice Center (AfricaRice), Cotonou, Benin; E-mail: j.d.reece@cgiar.org*

Rubyogo, Jean Claude, *International Center for Tropical Agriculture (CIAT), Lilongwe, Malawi; E-mail: j.c.rubyogo@cgiar.org*

Silué, Drissa, *c/o World Vegetable Center (AVRDC), Arusha, Tanzania; E-mail: silue.drissa@gmail.com*

Sperling, Louise, *International Center for Tropical Agriculture (CIAT), Arusha, Tanzania; E-mail: l.sperling@cgiar.org*

Ugen, Michael A., *National Agricultural Research Organization (NARO), Namulonge, Uganda; E-mail: michaelugen@yahoo.com*
Van Mele, Paul, *Africa Rice Center (AfricaRice), currently Agro-Insight, Ghent, Belgium; E-mail: paul@agroinsight.com*
vom Brocke, Kirsten, *International Crops Research Institute for the Semi-Arid Tropics (ICRISAT), Bamako, Mali; E-mail: k.vbrocke@icrisatml.org*
Wanyama, David, *National Seed Certification Service (NCRS), Entebbe, Uganda; E-mail: dnwanyama@yahoo.com*
Weltzien, Eva, *International Crops Research Institute for the Semi-Arid Tropics (ICRISAT), Bamako, Mali; E-mail: e.weltzien@icrisatml.org*

Foreword

The world is currently struggling to harness its resources to meet the growing challenges of food production in the face of population growth, climate change and increased prices of energy and other production factors. The second World Seed Conference (Rome, 8–9 September 2009) revealed that the global seed market has grown rapidly in recent years and is currently around US$37 billion. However, developing countries are yet to be part of the major seed trade. The conference therefore called for urgent government measures to increase public and private investments in the seed sector for agriculture to meet the challenge of food security.

It is gratifying to note that, in all African countries, some arrangements exist to provide seeds for farmers, attesting to the recognition by African governments that this input is critical for food security. However, in spite of good intentions, seed programmes often do not or no longer meet the needs of farmers. In southern and eastern Africa, private sector participation is more advanced, resulting in relatively better developed seed industries, even if only for a few selected and commercially attractive crops. In the rest of Africa, the rapid pace of seed sector development in the past seems not to have resulted in sustainable programmes, as most of it is concentrated on state farms, parastatals and other publicly funded and operated seed activities across the board. Economically they were not sustainable as public budgets dwindled and donor support waned.

The recent trend is for the public sector to disengage from seed production, although responsibility for early generation seed production continues to be entrusted to some public funded agencies.

As the public sector withdraws, the degree of abandonment of seeds of crops (self- and cross-pollinated) important to food security but of little interest to multinational companies becomes even more pronounced and that of vegetatively propagated crops such as cassava, plantains and yams even direr. Indeed, this is the present situation in most countries where seed development is following modern trends. Clearly some imaginative ways have to be found to enable a sustainable seed production and distribution effort to take off for food security crops in Africa.

As an initial step, FAO and AfricaRice joined efforts to document, throughout Africa, success stories of small and medium-sized seed enterprises dealing with non-hybrid food crops and identify the factors behind their success.

The case studies from nine African countries presented in this book provide an overview of the current status of the African seed sector, and associated challenges and opportunities. They form a solid basis for a broad-based discussion on how to strengthen the sector to the benefit of food security across the continent.

The studies showed that the challenges facing the seed sector in Africa are many and varied including at institutional and infrastructure levels. However, many African seed enterprises are able to bring together the formal and informal seed sectors in rather innovative ways, thereby assuring a flow of improved seeds to small farmers. The sustainable development of these enterprises is challenging in most countries due to the absence of an enabling environment. Entrepreneurs face non-adapted legislation, poor infrastructure, lack of appropriate financial products and business support services and lack of training. The extent to which each country is able to address these challenges and empower all the major actors will undoubtedly positively impact the growth of the seed sector.

The lessons learnt from these studies should help support the rapid development of the African seed sector and increase the steady supply of quality seed of food security crops to millions of African smallholder farmers.

Modibo T. Traoré
Assistant Director-General
Agriculture and Consumer Protection Department
Food and Agriculture Organization of the United Nations

Papa A. Seck
Director General
AfricaRice

Introduction to this Book

The subject of seed provision commands an exceptional amount of attention in most discussions of agricultural development. The reasons are not difficult to understand, as the security and quality of seed supply are among the principal determinants of any farmer's success. But, despite this unanimity of interest, there is relatively little agreement on what needs to be done to support the growth of effective and equitable seed provision in developing countries.

Part of the controversy over seed provision is the product of legitimate concerns about the nature and impact of the commercial seed sector. There is, for instance, the danger that an overemphasis on commercial seed supply will disregard the role that farmers continue to play in the identification and preservation of productive germplasm, the maintenance of local systems of seed sale and exchange, and innovations in crop management. In addition, there are understandable worries about the growing reach of the multinational 'life sciences' industry, its increasing control of the seed sector and the concentration of access to technology in too few hands.

But commercial seed supply can take many different forms, and it is a larger part of many 'traditional' farming systems than people may realize. It is difficult to imagine a productive agricultural system in the 21st century without access to some type of formal seed provision through various types of seed enterprise. As agricultural economies develop there is a natural shift towards specialization in the supply of products and services that were previously part of self-sufficient farms or communities. The access to commercial seed supply offers wider access to the products of modern plant breeding and helps ensure that a farmer's seed supply is not completely dependent on the vagaries of local climate or other uncertainties in local production systems. In addition, the expansion of agricultural markets often provides farmers with an opportunity to earn a premium for the specific qualities of their produce, which may require more attention to high quality and uniform seed than can be provided on-farm. The availability of commercial seed also allows the farmer to invest time in other activities, on- or off-farm, without having to worry about next year's seed supply.

Seed is a special product, embodying specific genetic information as well as important physical qualities, and as such its commerce offers particular challenges and opportunities. Some of the challenges are related to the fact that many of the seed's qualities are not immediately obvious to the buyer and hence some type of regulatory system may be required. But, even where seed regulatory regimes are in place, they are rarely successful unless there is a concomitant exercise of farmers' own capacities to understand input markets and to learn about the products on offer, and the investment of seed suppliers in establishing and promoting their reputations in the farming community. Formal seed provision thus offers a stimulus to the expansion of robust agricultural markets by developing the trust and knowledge that allows participants to make productive exchanges.

The expansion of formal seed provision also offers other contributions to the development of the agricultural economy. The inability of public plant breeding programmes to deliver many of their varieties to farmers has been a chronic problem. The emergence of local seed enterprises (many of which will depend, at least initially, on public varieties) offers a solution to this dilemma, places increased pressure on public research to deliver products that farmers can use and offers opportunities for productive public–private interactions that are crucial to agricultural development. Although there are some economies of scale in seed production, experience has shown that relatively small seed companies can be viable, thus offering the opportunity to contribute to the development of small and medium-size businesses supporting local agriculture. In addition, the fact that most commercial seed production is done through contracted seed growers offers an opportunity for some farmers to increase their earnings through specialized production.

There has been much investment by donors and national governments over the past several decades in support of the formal seed sector in developing countries. Some of those investments are beginning to pay off, but there is much that remains to be done. This is particularly true in sub-Saharan Africa, where there are many challenges to establishing viable businesses. Although the literature on seed systems includes various attempts to describe patterns of seed enterprise growth and to delineate distinct stages in seed industry development, these tend to be oversimplified approximations. The reality is much more complex and idiosyncratic. The performance and prospects of seed enterprises depend on a wide range of factors, including the specific crops, the characteristics of national agricultural economies and the adequacy of local institutions.

There is thus much to be learned about what is needed to ensure seed enterprise development. That is why *African Seed Enterprises* is important, because this book describes an exceptionally wide range of experiences in sub-Saharan African countries. Each of the cases is substantially unique and presents a particular history and set of lessons. The descriptions are accompanied by country profiles that allow the reader to appreciate the agricultural and policy environment that determines the trajectory of these enterprises. Not all of the enterprises described in this book are likely to survive, and some are still perilously small operations or highly dependent on external support. But their experience to date offers an exceptionally rich source of data and ideas for understanding seed systems and appreciating the diversity of strategies and resources that have been used to create new seed enterprises. These are the varied and often imperfectly formed building blocks from which viable commercial seed systems are being constructed. The experiences should be of great use to governments and donors when they consider how to provide more adequate support to seed enterprise development.

Robert Tripp

Acknowledgements

We thank Marco Wopereis for encouraging this study and for reading an earlier version of this book; AfricaRice for allowing staff to work on this study and for providing office space and logistical support; FAO for initiating and supporting this study. Additional resources were provided by USAID and IFAD. The Netherlands Organization for International Cooperation in Higher Education (NUFFIC) partly funded the research in Guinea, as part of a PhD programme.

We thank all the national programmes and seed entrepreneurs featured in this book who shared their valuable time and insights.

We thank Abdoulaye Kabore, Louis Béavogui, Malick Soumah, Moustapha Ceesay and N'Famoussa Camara for their assistance in collecting additional data.

Aliou Diagne provided input for Chapter 2. Abderrahmane Lyamani, Cathelijne van Melle, Daniel Dalohoun, David Wanyama, Dinah Borus, Djoukamady Diallo, Drissa Silué, Essa Drammeh, Florent Okry, Jean Claude Rubyogo, Judith Oggema, David Reece, Kehinde Adelugba, Marco Wopereis, Michael Misiko, Olupomi Ajayi, Sékou Diawara and Tom Remington contributed ideas to the concluding chapter.

Conny Almekinders and Graham Thiele kindly provided constructive criticism on an earlier version of this book.

Other people who reviewed one or more chapters are Clive Drew, Louise Sperling, Marco Wopereis, Peter Matlon, Solveig Danielsen and Tom Remington.

List of Tables

List of Boxes

1 Introduction: a Full Granary

ROBERT G. GUÉI, JEFFERY W. BENTLEY
AND PAUL VAN MELE

1.1 Why We Study Success Stories

A granary full of cereal can start with a single seed, the most renewable agricultural resource. But seed is also alive and delicate, and in most developing countries farmers still cannot get enough good quality seed to make their countries food secure. Farmers often can only get quality seed if there are viable seed enterprises that multiply, store and distribute seeds. These enterprises can be small family farms, public agencies and medium-sized companies. This book takes a people-centred look at those enterprises, at the people with the skills, talent and gumption to open the doors of a business as difficult as seed, and to keep it running in a place as challenging as Africa.

Seed is often expensive, placing poor farmers at a disadvantage. Large seed companies concentrate more on countries with big farmers and a large demand for seed, especially hybrid maize and vegetable seed. They often ignore seeds with thin profit margins, such as self-pollinated crops, like wheat, rice and beans, open-pollinated crops like non-hybrid maize or vegetatively propagated crops, because farmers often save the seeds from one harvest to the next and because proprietary laws are missing or not enforced. However, these are the crops still widely grown by most smallholder farmers, providing food and employment for themselves and others. Letting farmers try new varieties and then distributing the seed will be crucial as the world adapts to climate change.

Nevertheless, there are a few unsung examples in Africa and elsewhere of successful seed enterprises (see Box 1.1) running on less equipment and modest capital investment. It is important to identify and describe their unique experiences in the search to build durable, quality seed systems in Africa.

Recent case studies by FAO and partners in Brazil, India, Côte d'Ivoire and nine other African countries aim to find the constraints and the keys to overcoming them to build successful seed enterprises. We hope that the business models and experiences will inspire others to produce and distribute seed for Africa's food security crops. The seed for these crops is less profitable than vegetable seed or hybrid maize, but it is possible to make money and serve the community at the same time, as the cases in this book show.

> **Box 1.1 Definition.**
>
> ***Successful seed enterprise***. Any farmer, association, small or medium-sized company, public agency or any other person or group that is able to stay in business for several years, producing or selling seed, overcoming the cash flow challenges of seed. The enterprise keeps its customers or finds new ones by offering quality seed and by creating new products and services. The enterprise should be able to manage and finance itself.

1.2 The Issues

Specializing in seed is difficult. Farmers must have the seed ready at planting time, which means they must either grow it off-season, save it from the previous harvest or bring it from somewhere else. They must invest time and money months before their product is ready to sell, only to find that by planting time many of the seed customers are out of money.

The seed must be of high quality. Farmers disappointed by poor germination or by seed-borne disease lose confidence in the seed producers. Seed has to be planted and tended to know if it is any good (Tripp, 2001). So certification, or brand names, became a way of signalling quality, but the certifier and the enterprise must keep a consistent high quality.

To grow healthy seed, a crop must be in excellent condition. The field needs to be kept especially free of weeds and diseases to avoid passing health problems on to the next generation. Formal seed is often dusted with a coating of fungicide as an added protection. Then the seed must be packaged, labelled and marketed.

In spite of these problems with cash flow and quality, in the 1980s some development workers began to see an inherent logic in encouraging smallholder farmers to produce the seed for common food crops, especially self-pollinated (like rice, finger millet, beans, cowpea and groundnuts) and cross-pollinated crops like maize, sorghum and pearl millet. After all, the farmers were already producing and managing seed for their own crops, often selling or giving seed to their neighbours as well. With a little effort they could grow seed to sell as well. Plus the farmers as seed producers were close (geographically and culturally) to their potential clients. And there was the appeal of giving relatively low-income people a new money earner, while promoting new, high yielding or disease resistant crop varieties.

But it wasn't as simple as that. Producing seed for sale at a larger scale requires many steps which are omitted in farm-saved seed. National seed regulations often inhibit trade by unregistered entrepreneurs or trade of local varieties that are not included in the national seed catalogue (De Schutter, 2009). Those interested in becoming a seed producer must have ties with research and plant breeders to get the 'foundation seed' needed to grow certified seed. Then the producers must contact a perhaps distant government seed certification agency and arrange for inspections while the seed crop is in the field. This costs money and requires management skills and contacts. Turning village farmers into a group of organized seed producers requires that they work together, from seed to sale, while often the farmers find it easier to do most tasks alone or just with household members.

Government seed certification agencies tend to have limited staff and slim travel budgets and cannot inspect seed plots of hundreds of scattered smallholder seed producers. But, if outgrowers work as a group and sow their crop at the same time, all plots in a given area can be inspected on the same day. However, organizing farmers is easier said than done. They find it easier to take decisions and organize their work at the household level (Netting, 1993).

Producing and distributing high quality seed often require investment in labour, machinery, chemicals, processing and storage facilities, packaging and marketing. Investors only embark on it if they are sure of having a return to their investment. After independence, African countries understood that in order to feed their populations agricultural development should go hand in hand with the development of their seed sector.

In the absence of a private sector, governments invested heavily in producing seed of most of their food security crops, most of which are either self-pollinated, open-pollinated or vegetatively propagated. Although some were successful, the structural adjustment policy in the 1980s forced governments to abandon seed production and distribution without considering the difficulties for the private sector to assume responsibility for producing the less profitable seeds of food security crops.

Much of the equipment in the seed plants that survived in the public sector is now 30 years old, over-sized, and needs replacing.

Many countries have encouraged privatization and withdrawn from investment in plant breeding and seed production. Over 90% of the crops in developing countries are grown from farm-saved seed (Minot *et al.*, 2007), although much of the saved seed is of modern, high-yielding varieties. However, on the whole, private businesses did not come forward to fill the gap left by government withdrawal. Donors and NGOs started to support farmer seed production.

After food prices spiked in 2008, African governments became concerned about increasing their food production, in part to lower the costs of importing food, especially rice. Yet African farmers who were planting modern crop varieties were reaping harvests of up to double those of their neighbours. And governments complained that, years after national and international centres had released new varieties, the seeds were still in short supply. By 2010, African governments, the FAO and other donors were once again emphasizing improved seed.

Global changes. In 2050 the world's population will be 9.1 billion, 34% higher than in 2010. All population increase will occur in developing countries, particularly in Africa. Seventy per cent of the world's population will then be urban, compared with just 49% today (FAO, 2009). Food production must increase by 70% to feed this larger, more urbanized and wealthier population (which will buy more meat, milk and eggs). The world will need genetic diversity to adapt to global warming. As farmers respond to improved transportation and opportunities to sell new crops, from fruits and vegetables to biofuels, they will need a more constant access to seeds of new crops and varieties. Changes in world market prices also affect choices made by the local agroprocessing industry. Facing rising costs of barley, for instance, breweries in several African countries have shifted to sourcing locally grown sorghum, thus immediately influencing seed demand.

Strong research programmes and solid seed systems for all crops are crucial, particularly in Africa, to give farmers access to better varieties and quality seeds at all times, in all sites and for all cropping systems and markets.

A study commissioned by the Rockefeller Foundation on seed development programmes in sub-Saharan Africa describes issues of varietal development, seed multiplication, marketing, output markets and capacity building for major food security crops (Minot *et al.*, 2007). Most seed studies, however, compare cases and draw conclusions without presenting an in-depth analysis of the seed enterprises themselves. Hence FAO and partners decided to systematically document a wide range of successful enterprises that produce or distribute seed of local food crops to respond to farmers' demands. These workable business models will hopefully inspire others.

1.3 Case Studies in Brazil, India and Côte d'Ivoire

In 2008 the FAO commissioned studies of successful seed enterprises in Brazil, India and Côte d'Ivoire (FAO, 2010).

Brazil has successful seed enterprises for soybean and open-pollinated maize. Soybean is a commercial crop grown on farms ranging from 10 to 100 hectares in the south and 500 to 5000 hectares in central Brazil. The soybean case study looked at a private seed production enterprise (Maua Seed Company) established in Paraná State (southern Brazil) in 1975 by three agronomists who obtained varieties under licence from Embrapa (the national agricultural research agency) and helped farmers get loans to produce seed. As more farmers joined in, seed production increased from 2800 t in the 1970s to about 20,000 t in 2008.

In contrast, farmers growing non-hybrid maize are mainly smallholder subsistence farmers. The maize seed case study examined a collaborative enterprise established in 1999 by farmer groups with financial support from an NGO (AS-APTA), foreign donors and government and technical support from the State University of Londrina. The programme started working with 1500 organized farmers in ten towns in Paraná. By 2006 the programme had expanded to 4000 farmers in 22 towns. Seed is sold through local agricultural retailers and at local community fairs.

In India, the public sector supplies predominantly self-pollinated varieties and the private sector supplies hybrids. However, private sector participation in sorghum, pearl millet and rice seed production has been encouraged by the low marginal cost of getting into the business and the high seed turnover rate.

Also the cooperative system is highly developed in India, often supporting farmers with credit, inputs, storage, processing and marketing facilities. One of the most successful cases is the Mulukanoor cooperative in Andhra Pradesh. The agricultural officers employed by this cooperative monitor seed crops in the field and train the farmers in quality control. The cooperative buys back all seed produced, after deducting the cost of inputs and credit advanced. Seed is processed in the cooperative's own seed processing factory before it is sold to various seed companies in the states of Andhra Pradesh, Karnataka and Maharashtra for commercial paddy production.

In Côte d'Ivoire a true public–private partnership in variety development and seed production was the basis for success. The new rice varieties were bred by public research institutions, seeds produced by a large rice cooperative and the quality control done by the national seed service at minimum cost. Linking production of quality seed to paddy production sustained the demand for seed. The cooperative provided training in rice and seed production for participating farmers.

Key lessons. From these three studies the FAO learned that small-scale seed enterprises are nurtured by: government support, access to credit, entrepreneurship, technical skills and capacity, sustaining demand for quality seeds, enterprise ownership and profitability, adequate equipment, improved varieties and early generation seeds, and links between formal and informal seed sectors.

1.4 Method

Based on the lessons from the first three case studies and given the need to support seed for the long-term development of African agriculture, in 2009 the FAO asked AfricaRice to conduct a study of successful seed enterprises in Africa. The cases are quite diverse, including public agencies in Morocco and Nigeria, private companies in Madagascar, Kenya, Uganda, Nigeria and elsewhere, farmer groups in countries like Mali, Cameroon and Guinea, and enterprising farmers in The Gambia, Guinea and Kenya. The FAO and AfricaRice wanted to know how these enterprises had survived, and often prospered, at such a challenging task as producing a delicate product with specific cash flow challenges and slim profit margins for a scattered and often cash-strapped market.

Potential case studies of successful seed enterprises were identified after consulting with national and international experts. We selected a few small and medium-size enterprises with at least 5 years of successful operation, which were producing self-pollinated and open-pollinated crop seed. However, this seed may just be one activity. A seed enterprise may also be producing hybrid maize seed, vegetables, or importing vegetable seed in order to survive, while also selling self-pollinated crop seed.

Country chapters were written based on past experience of the authors. AfricaRice social or agricultural scientists with experience in Africa and seed, and partners in each of the countries, visited the enterprises, interviewed the managers and other staff members, visited the facilities, the outgrowers, traders and credit providers and talked to researchers as well as other people who know the seed sector.

All the studies were reviewed by the authors themselves during a regional workshop, as well as by outside peer reviewers (experts in seed systems), and were edited for clarity and to give the book a uniform voice.

1.5 How the Book is Organized

The book opens with a review of how seed enterprises work (Chapter 2) followed by nine country chapters and a conclusion. Each country chapter is written by AfricaRice staff, FAO staff, one or more national collaborators (technical people familiar with local seed activities) who helped plan and guide the work, and occasionally staff from CIAT, CIP, ICRISAT and Wageningen University.

The chapters are written as case studies of successful enterprises. We discuss a few of the mistakes the enterprises have made, but we omitted cases which we deemed unsuccessful. The country chapters are arranged geographically, starting with Cameroon, and proceeding clockwise around Africa through Nigeria, Mali, Guinea, The Gambia, Morocco, Kenya, Uganda and Madagascar. The book ends with lessons learned and recommendations for development agencies, donors, decision makers and banks.

Country context. Each chapter starts with an overview of the country's agriculture (farms, crops, agricultural economy and history of agriculture). It then describes the country's seed sector, seed policies and norms, and the influences on seed supply and demand.

Next, each enterprise in the study is described individually, covering history, structure, cash flow and marketing.

History includes what type of enterprise it is (e.g. family farm, cooperative, private company), its size, why the founders were motivated to get into the seed business, how the whole enterprise fits together, how it evolved and the events that shaped it.

Structure is a study of how the seed enterprise is put together, including management, workers, land, equipment, suppliers and customers, activities, links with research and extension, and seed quality control. Management's vision, how they see the future and adapt to change, as well as the technical capacities of the enterprise, is described. The section discusses growing seed off-season or using storage to save seed to sell at the main planting season.

Some of the enterprises buy seed, while others grow it, but most use outgrowers. The section describes which category of seed is sold (breeder seed, foundation seed or certified, etc.), and the main activities, e.g. other goods or services they provide, besides seed.

How the enterprise perceives national seed policies and laws and how it deals with plant variety protection (PVP) issues and uses public resources, such as capital or training, are discussed.

Cash flow. This section assesses how the enterprise manages to obtain inputs for seed production and then wait for the next growing season to be able to market seeds. Few of the enterprises have easy access to loans. Ironically, many seed enterprises are forced to provide credit, since sometimes farmers, distributors or agro-dealers have no cash to buy seed at planting time.

Marketing strategy and how the enterprise estimates seed demand and plans seed production are described. Some enterprises sell mostly to farmers while others sell to projects, international organizations, government agencies, retailers or NGOs. All successful enterprises manage to keep customers and maintain their trust.

Farmers often buy seed. Understanding the importance of direct sales to farmers and others sheds light on the vision of seed enterprises.

All seed enterprises have to maintain a name brand. Some do little advertising and a few have explicit links to the media. Some use field days and other direct-to-farmer forms of advertising. The enterprise must have a way of learning about changing demand and meeting it.

Conclusion. The chapter ends with a conclusion on the challenges and strengths of the enterprises in the country, including their limitations and constraints, but stressing the strategies they use to overcome those, i.e. the main reasons for the success and the sustainability of the enterprises, and their major strengths.

1.6 Audience

We avoided technical jargon when possible in order to appeal to a wider audience. Topics are discussed that matter to seed enterprises, that are often discussed by people in the seed sector, but that are rarely documented. The book hopes to inspire future generations of researchers, development workers and entrepreneurs.

But mainly we address the decision makers who set policy, plan research and manage breeder and foundation seed, as well as donors and development agencies supporting seed development in Africa. Last but not least, we hope that people who create and manage seed enterprises will find the book useful.

References

De Schutter, O. (2009) *Seed Policies and the Right to Food: Enhancing Agrobiodiversity and Encouraging Innovation.* Report to the United Nations General Assembly (A/64/170). United Nations, New York. 22 pp.

FAO (2009) High Level Expert Forum. How to Feed the World in 2050. Rome, 12–13 October 2009. http://www.fao.org/fileadmin/templates/wsfs/docs/Issues_papers/HLEF2050_Global_Agriculture.pdf.

FAO (2010) Promoting the Growth and Development of Smallholder Seed Enterprises for Food Security Crops: Case Studies from Brazil, Côte d'Ivoire and India. AGP Series 201. FAO, Rome. 34 pp.

Minot, N., Smale, M., Eicher, C., Jayne, T., Kling, J., Horna, D. and Myers, R. (2007) *Seed Development Programs in sub-Saharan Africa: a Review of Experiences.* Rockefeller Foundation, Nairobi. 189 pp.

Netting, R.M. (1993) *Smallholders, Householders: Farm Families and the Ecology of Intensive, Sustainable Agriculture.* Stanford University Press, Stanford. 389 pp.

Tripp, R. (2001) *Seed Provision and Agricultural Development: the Institutions of Rural Change.* Overseas Development Institute, London. 174 pp.

2 How Seed Works

JEFFERY W. BENTLEY, PAUL VAN MELE
AND J. DAVID REECE

The purpose of this chapter is to compare formal and farmer seed systems, to describe the role of quality seed and to present the context in which small and medium-sized enterprises work in Africa, with a particular emphasis on the challenges of farmer seed enterprises. The demand for seed varies widely with the type of crop planted, and this influences which types of seed farmers want to buy. One of the most important reasons to buy formal seed is to acquire a new variety. The formal sector often ignores these reasons, and complains of how little seed farmers buy, especially smallholders in Africa and elsewhere in the tropics. Yet farmers can often obtain quite good seed from markets, their neighbours or their own farms. The formal seed system is actually a marriage between the government and the private sector, a marriage where communication and collaboration happen some of the time, but not always. While some have proposed encouraging groups of smallholders to produce formal seed for others, this has not been a success. Smallholder seed producer groups find it even more difficult than private seed companies to collaborate with the public sector for source seed supply and certification services. In light of these problems, the cases of functional African seed enterprises described in Chapters 3 to 11 seem even more remarkable.

2.1 Formal and Informal Systems

Farmers get seed from 'formal' or 'informal' systems (Sperling *et al.*, 1996; Sperling and Cooper, 2003). Formal systems are purposively composed of separate activities to provide new varieties, maintain their purity, certify the seeds and distribute them to farmers, usually through officially recognized seed outlets. De Schutter (2009) labels the formal sector the 'commercial seed system', which is not quite accurate, since there is much public sector involvement in it as well.

An 'informal', 'traditional' or 'farmer' seed system lacks public sector regulation (Thiele, 1999). Farmers frequently exchange seeds among themselves, often for sale (Almekinders *et al.*, 1994; Almekinders and Louwaars, 1999; De Schutter, 2009). Farmer seed systems also develop new varieties and maintain crop genetic diversity, but they do it as an integrated part of crop production (Almekinders and Louwaars, 2002; Brush, 2004).

The stated goal of formal systems is to maintain quality of seed, but high quality may be maintained even without legal recognition, e.g. in the UK, where seed potatoes no longer need be certified, or in the USA, where most hybrid maize seed is not certified, and yet in both cases the quality is high. There is a movement in industrialized countries towards quality declared seed (QDS), where quality is maintained

Table 2.1. Formal vs informal seed systems.

	Formal system	Informal (popular or farmer) system
Goals	Distribute high quality seed of modern, high yielding varieties	Obtain seed to sustain the farm every season
Quality of seed	Variable, but usually high	Variable, but usually high for most crops
Public sector	Source seed, research and certification	Not necessarily involved
Private sector	Multiplication and distribution by registered enterprises	All activities. Distribution by farmers, registered or unregistered traders and vendors
Seed type	Formal, certified seed	Common or farm-saved seed

without regulation (Graham Thiele, personal communication). Quality problems occasionally creep into formal seed systems, so there is not a simplistic correspondence between 'formal' seed and high quality (Table 2.1).

2.2 Demand for Seed

Farmers need a good reason to buy seed (Tripp, 2003). But there are many reasons and at least 20–30% of the non-commercial seed planted by African farmers is from off-farm (Tripp and Rohrbach, 2001).

Poverty. The poorest may be tempted to sell their entire crop at harvest, or to eat it, without saving seed. Many poor bean farmers in the Great Lakes region of Africa buy all their seed. A third of poor farmers surveyed in Rwanda bought all of their seed, while in Burundi and Zaire (DR Congo) 70% and 52% did so (Sperling *et al.*, 1996). Fresh seed is frequently sought to make up for a poor harvest, after seed stocks have been eaten or sold (Tripp and Rohrbach, 2001).

Quality problems. Some seed deteriorates after several generations. For example, viruses build up in seed potato, especially at lower altitudes. Storage conditions (too dark, too humid, too dry, tuber moths, etc.) may reduce seed quality. The rate at which planting materials deteriorate and need to be replaced varies between crops and regions and also with the skills with which farmers manage and store their crops. Farmers make economically rational decisions to renew seed, i.e. they must decide if the cost of seed replacement is compensated by the added value from the new seed. The botanical facts of life dictate that some seed must be renewed frequently to maintain quality (e.g. hybrid maize, potato in the lowland tropics), while farmers can replant some seed virtually indefinitely (e.g. self-pollinated rice or wheat varieties) with little or no loss of quality (Table 2.2).

Table 2.2. Important biological features of some major crop types. (Adapted from Lanteri and Quagliotti, 1997.)

	Hybrid maize	Open pollinated maize	Sorghum millet	Wheat	Rice	Beans	Potato	Groundnut
Breeding system	Forced out-pollination	Aggressive cross pollination	Cross pollination	Self pollination	Self pollination	Self pollination	Vegetative propagation	Self pollination
Sowing rate per ha	Medium (20–35 kg)	Medium (20–35 kg)	Low (10–20 kg)	High (150–200 kg)	High (70–150 kg)	High (100 kg)	High (800–2000 kg)	High (125 kg)
Multiplication factor	High	High	High	Low	Medium	Medium	Very low	Very low
Rate of deterioration	Very rapid	Potentially rapid	Medium	Slow	Slow	Very slow	Rapid at low altitudes	Very slow
Type of deterioration	Genetic (increased homozygosity)	Seed acquires genes from other maize varieties unless field is isolated	Acquires genes from other varieties if field is not isolated	Pollination occurs within the plant, creating genetic stability	See wheat	Like wheat only more so	Genetically stable, but disease loads build up faster in seed at lower altitudes	
Frequency of purchase (average)	Annual	2–3 years	3 years	4 years	4 years	Variable	Variable	Variable

New varieties. Farmers often buy seed to obtain a new variety, but, once they have the variety, farmers may be reluctant to spend money on the seed again (Sperling *et al.*, 1996). For example, in eastern Kenya 67% of farmers who adopted NPP 670, a modern variety of pigeon pea, bought the seed when first planting it, but only 38% acquired seed more than once (Jones *et al.*, 2001). A study of the adoption of high yielding maize and wheat varieties in East Africa found high levels of adoption (75 or 90%) among smallholders in higher potential regions. However, after the farmers' first experience with the variety, the use of certified or purchased seed was low (Doss *et al.*, 2003).

Crop characteristics. Table 2.2 illustrates variation in seed demand by crop. There are other crop types and other facts of life not illustrated in the table, and the features vary depending on the system (e.g. commercial potatoes in Europe, versus East Africa or subsistence farming in the Andes). But the point is that the crop one plants influences how often one seeks seed.

Vegetables. Some vegetable seeds may be too tedious to produce on a small farm, or the preferred varieties may be easier to grow in Europe or some other temperate climate. So it may be convenient for tropical smallholders to buy vegetable or luceme seed every season.

Low seed requirements. To plant a field with coarse grains farmers require low amounts of seed so they can more easily afford to buy it (Tripp, 2003). If one is planting a hectare of maize or sorghum with 20 to 30 kg of seed, the added cost of certified seed may be affordable. But seed potato is so bulky it may take 2 tonnes to plant a hectare. The cost of certified seed would be such a big expense for potato growers that they prefer buying common seed, rather than certified (see Bentley and Vasques, 1998, for an example from Bolivia).

Hybrid maize is attractive to private seed companies because yields decline and the crop is no longer uniform if the grains are used as seed, so farmers tend to buy new seed every season. Seed companies first promoted hybrid maize seed in the United States in the 1930s as a way of forcing farmers to buy seed. Before then, companies had sold mostly vegetable seed (Kloppenburg, 1988).

The development of high yielding maize in Africa has transformed maize from a minor crop in 1900 to the largest source of calories today. Breeding was started in Zimbabwe (then Rhodesia) in the 1930s and in Kenya in the 1950s. Both countries made breakthroughs with hybrid maize in the 1960s. At independence, African countries targeted smallholders, with major surges in productivity related to marketing and input support, including seed enterprises to supply the seed (Gabre-Madhin and Haggblade, 2004).

Farmers may be reluctant to buy hybrid maize seed because its price is seven to ten times higher than the grain price. This is why only a handful of private companies produce it and multinationals have pulled out of several African countries after having tried to sell hybrid maize seed. Most rural retailers survive on the basis of rapid turnover of limited capital obtained through the sale of commodities for which demand is better known, such as vegetable seed and pesticides. These traders are reluctant to risk their scarce capital stocking commodities such as seed of cereals or legumes, for which demand is less certain (Tripp and Rohrbach, 2001).

The next section suggests that farmers can often get good seed from informal sources.

2.3 Supply of Seed

The UN recognizes that 'States are obliged to protect their citizens' right to food, which applies both to the regulation of commercial seed systems and to the preservation and enhancement of informal or traditional farmers' seed systems' (De Schutter, 2009). Although commercial seed varieties may improve yields in the short term, their higher performance often has been a response to inputs (fertilizers) and to water availability, making it difficult for farmers unable to access such inputs and conditions to reap their benefits. Therefore states must also support farmers' seed systems (De Schutter, 2009).

Farmers tend to use both formal and informal seed systems, and the same varieties may move back and forth from both. For example, breeding material may originate from landraces, and farmers may easily appropriate modern varieties (Sperling and Cooper, 2003).

2.3.1 The supply of common seed

Farmers get seed from various sources.

Farm-saved seed. The cheapest and closest source of seed is one's own farm. A survey of maize farmers in north-west Nigeria found that 71 to 100% of their land was in improved varieties, although most were using farm-saved seed (Daniel and Adetumbi, 2006). In south-west Nigeria improved vegetables are grown widely but '60% of vegetable farmers save their seed' (Daniel and Adetumbi, 2004). Farmers may well be attracted to new varieties but not buy the seed every season.

Farmer to farmer. Seed exchange networks are pervasive and yet patchy. Tripp and Rohrbach (2001) found that most smallholder farmers either give or receive seed each year, through gifts or barter and also sales. However, networks link relatives and friends more closely than neighbours (Marfo *et al.*, 2008). Seed networks may be limited by class barriers, while migration and increased market production tend to fragment seed exchange networks (Almekinders *et al.*, 2007). Commercial transactions are replacing seed exchange and gift-giving among farmers (Rubyogo *et al.*, 2010). However, smallholder farmers do not necessarily construct special seed systems; they may simply use their existing networks to get hold of seed (Badstue, 2006).

Farmer seed systems can distribute new crops quickly. In a community in Sulawesi cacao growing started in the 1980s when a woman was given seed by someone from outside the community. She planted it as an ornamental. She threw away the seeds for years, and was thrilled in the late 1990s to learn that they could be sold. Before 1990 hers was the only household that grew cacao; by 1996 75% did and by 1999 100% did. This rapid change and the spread of the cacao seed occurred without any outside support (Belsky and Siebert, 2004).

Farmer-to-farmer seed exchange may be for cash. When an attractive rice variety reached Ghana from neighbouring Côte d'Ivoire, two-thirds (67%) of farmers bought the seed from another farmer, while others received the seed as a gift or bought it in the market, while seed exchange was unimportant, just 2% (Marfo *et al.*, 2008).

Local markets sell grain and seed, including local and improved varieties. Vendors include farmers selling their own seed, wholesalers and retailers. Seed trade in markets varies considerably, but is an important source of seeds for low-income farmers (Lipper *et al.*, 2009). When traders in Somalia invest in good seed and storage, farmers prefer buying seed from them in local markets (Longley *et al.*, 2001). Local markets are important sources of seed, especially in times of emergencies when farmers may run out of farm-saved seed (Sperling and Cooper, 2003). Many of the poorest farmers buy all of their seed from markets (Sperling *et al.*, 1996).

Sometimes quality may be a problem in market seed. A study of seed potato sold in Kenyan markets found only 27 out of 1000 tubers were virus free. Bacterial wilt, a serious potato disease transmitted in seed tubers, was found in 74% of potato farms. Of farmers interviewed in Kenya, Tanzania and Ethiopia, fewer than 5% got their seed potato from specialized seed growers. Most seed potatoes are produced as a by-product of growing ware potatoes. Most farmers surveyed got seed from their own farms or from neighbours and renewed seed only every six or seven seasons. But the article concludes that current seed potato prices justify saving seed. Research centres are the only source of clean seed (Gildermacher *et al.*, 2009), so potato seed growers must be linked to research.

Retailers. These are farm supply shops (called stockists in East Africa) that carry certified seed and agrochemicals, and so they sell more to 'progressive' farmers. However, few farmers in developing countries can afford to buy only certified seed, so retailers supply only a fraction of the crop seed needed in their area (Sperling, 2002).

A study in East Africa found that maize and wheat seed was readily available in shops in Kenya and Tanzania, less so in Uganda and Ethiopia (Doss *et al.*, 2003). To accommodate farmers' demand retailers may sell non-certified seed of local or improved varieties if it comes from a trusted source.

Retailers face great risks, as farmers buy inputs within a narrow planting window, and, depending on farmers' forecasts of conditions and their disposable cash, stockists run the risk of being left with stock which cannot be kept for another year. Seed selling also requires more knowledge than retailing drinks, soap and groceries, which can turn over their capital regularly throughout the year (Poulton *et al.*, 2006).

2.3.2 The supply of formal seed

The public and private sectors collaborate to form formal seed systems. Although differences exist between countries, the public sector (government or international research centres) generally develops improved crop varieties. National governments regulate the sector, approving and registering new varieties, and may certify seed.

National and international research centres often provide the breeder or foundation seed for the industry, but inadequate production of source seed is a major constraint in sub-Saharan Africa. Sometimes public institutions produce, store and distribute seed, but governments are more likely (especially after structural adjustment in the 1990s) to allow private enterprises to multiply and distribute the seed (Tripp and Rohrbach, 2001; De Schutter, 2009).

Seed certification. Trust is important when buying seed, because the quality of seed cannot be judged until long after it is planted. Seed certification replaces trust in the seller with trust in a government seed certification agency. This is to help avoid fraudulent sale in the private seed sector (Tripp, 2001). Certification regulations provide minimum standards of genetic purity, physical purity and germination (Rohrbach *et al.*, 2003).

Seed certification is almost always government or semi-government regulated, but there have been interesting experiences with private seed certification agencies, like the ones in Peru which were set up with support from USAID and the government of Peru (Bentley *et al.*, 2001). So seed certification can be privatized, although there are few cases, perhaps none in Africa. Seed certification is external quality control, and may not even be necessary if companies are concerned about maintaining consumer faith in their brand name (Tripp, 2001). Unfortunately, some developing countries see certification as a kind of magic bullet and adopt a model of rigorous seed regulation when they have hardly any seed industry in place. Especially in Africa, some countries insist that all seed for food crops be certified (Tripp, 2003).

Certification of quality is a kind of information sharing between buyers. Fafchamps (2006) concludes that market institutions in Africa are not so different from what they are in developed economies. For example, business ethics and informal networks matter everywhere. But certification is important for creating trust and standards in business when some formal institutions are missing, such as external audits, banking regulations and consumer protection.

Seed companies. Several seed companies may locate in the same area, often near a seed certification agency or some other facility. Seed expert Robert Tripp argues that these neighbouring seed companies then establish networks and share information and resources and reduce transaction costs (Tripp, 2001, 2003).

Seed companies often start with high value seed, like vegetables, and add lower value ones after getting clients and machinery (Tripp, 2003). They may also start with foundation seed production, which gives good value and is less bulky to market than certified seed.

As Tripp (2003) explains, unless seed is difficult to save or there is a fairly constant offering of new varieties, it is not likely that a seed enterprise will be able to base its business solely on the provision of new varieties. This is a mistake that is sometimes made in promoting new seed enterprises. Initial farmer demand for a variety gives the impression that sufficient commercial incentive is available, but after several seasons, when many farmers are growing the variety (and save their own seed or can get seed from their neighbours), any commercial advantage evaporates. Established seed companies may provide a fairly steady stream of new varieties, but such a flow is often not present in an emerging seed industry. Seed production offers economies of scope. Once a seed company has experience with a few seed products it is fairly straightforward to expand its portfolio (Tripp, 2003). Although established companies may venture in producing seed of legumes, its demand is highly unpredictable and quantities produced are often relatively low, perhaps with the exception of groundnut.

Groundnut seed is bulky and delicate. High seeding rates make farmers sensitive to price. As farmers need to store more than 10% of their harvest as seed for the next season, cash needs in the family often result in them selling off the portion kept for seed. Groundnut firms have an advantage in selling seed because they have already invested in supervising contract farmers and have ties to research (Tripp, 2003).

Varietal protection. Some form of protection is necessary to provide incentives for plant breeders. The standard argument is that protecting plant breeders' rights encourages them to create more varieties, and stimulates industry to provide those varieties for farmers. However, results are mixed, successes are case-based and there is little hard evidence to support the standard argument as a general rule. Seed and plant breeding industries have developed in some countries without legal protection. Developing countries are now adopting overly stringent varietal protection laws (Tripp *et al.*, 2007).

All countries that belong to the WTO must provide some protection for intellectual property (Tripp *et al.*, 2007). The WTO has asked African countries to adopt the standards of the International Union for the Protection of New Varieties of Plants (UPOV), even though these place unrealistic restrictions on smallholder farmers' abilities to save and exchange seed (Rohrbach *et al.*, 2003; Tripp *et al.*, 2007).

Most major countries have signed the UPOV plant breeders' rights. The restrictive UPOV rules permit farmers to use farm-saved seed but only on their own holding. Individual signatory countries can modify the rules to make them more lenient for farmers, but few do. The UPOV's Convention on Biological Diversity has failed to generate sufficient benefits to fund the conservation of biodiversity, and has sometimes created insuperable obstacles to the access of both researchers and the bioindustry to genetic resources. An official UN report concludes that the Convention's 'access and benefit-sharing is not adequate for domesticated plant genetic resources' (De Schutter, 2009). The CGIAR system and some other public agencies have made their genetic resources common property, believing that a public pool is a better way to manage plant genetic resources (De Schutter, 2009).

Excessive protection of breeders' rights and patents may actually discourage innovation, because plant breeding is a cumulative process, based on earlier plant material (De Schutter, 2009).

Plant variety protection often makes little sense in developing countries. Kenya is considering tightening regulations to forbid informal trade in wheat seed among farmers. Yet there is only one private producer of wheat seed, a company with government ties. And the restrictions will be impossible to enforce on many smallholder farmers. An overly rigid varietal protection scheme may actually reduce the flexibility required to nurture an emerging commercial seed system, and enforcing it may be too costly or impossible (and politically damaging) (Tripp *et al.*, 2007).

Temporary monopoly privileges granted to plant breeders and patent-holders through the tools of intellectual property supposedly encourage research in plant breeding. However, the poorest farmers may become dependent on expensive inputs. The farmers' seed systems may be put in jeopardy, even though the farmers still need them for their economic independence and to manage pests, diseases and climate change. The spread of the commercial seed system may damage farmers' ability to retain varieties suited to marginal conditions and may lead to the loss of genetic material which should be preserved for plant breeding. Sub-Saharan Africa is especially dependent on farmer seed systems (De Schutter, 2009).

In 2006 the African Intellectual Property Organization (OAPI) made up of 16 mostly francophone West and Central African countries implemented a single plant variety protection system, where one application covers all countries. This may make sense if it allows varieties to be shared more widely between countries.

Countries in West, Eastern and Southern Africa are pursuing the regional harmonization of their seed laws and regulations. This harmonization will help in removing the bottleneck of variety release committees and as such reduce the time and costs involved in releasing good varieties. Common regulatory frameworks are also expected to reduce the costs of trading seed and encourage scale economies in seed production. As a result, commercial seed production is expected to expand, providing farmers with improved access to new varieties and stimulating productivity growth. This implies greater regional interdependence of seed supply (Rohrbach *et al.*, 2003).

2.4 Enterprises in Africa

2.4.1 From colonial to global market forces

The colonial governments did little to prepare an African entrepreneurial group. In 1950 only one Nigerian had graduated from university. Britain, France and Belgium pulled out of their African colonies in 1960 or soon thereafter. Many of them hoped to hand over administration to Africans, but to continue profiting from African business themselves. Most African countries came under autocratic rule, which led to big wasteful projects. In Ghana, for example, Kwame Nkrumah set up big, money-losing corporations based on the Soviet model and large mechanized farms that steadily decreased production (Meredith, 2006).

Many national governments retained the agricultural marketing boards that had been set up by colonial governments to earn revenue through monopoly purchasing of commodities. In 1981 the Malian government paid rice growers 63 francs for a kilo of rice that cost 80 francs to produce (Meredith, 2006).

The 1980s was a lost decade for Africa. In the 1960s import substitution via industry had been popular, but industrialization collapsed in the 1980s. Civil servants' salaries fell. Many managers left and informal economies grew. In 1979 Senegal became the first to take a structural adjustment loan from the International Monetary Fund (IMF), dictating devalued currency, an end to subsidies, reduced tariffs, fewer government jobs, privatization of state enterprises and fewer restrictions on foreign investment. In the 1980s 36 African governments signed up with the IMF (Meredith, 2006). Structural adjustment was not successful, even by the standards of *Economist* writer Robert Guest (2004). Many privatized parastatals were sold to cronies at low prices. National per capita incomes declined and debts mounted (Meredith, 2006). Countries with IMF loans had zero or negative growth and often collapsed into anarchy (Easterly, 2006).

By the end of the 1980s the World Bank was getting tired of Africa, and the people of Africa were tired of their governments. In 1989 only three countries had multiparty politics, namely Senegal, The Gambia and Botswana (Meredith, 2006).

Thirty per cent of Africa's population lives in landlocked, resource-scarce countries (Collier, 2008). Being landlocked makes it especially difficult to export agricultural commodities (which stimulates demand for seed), and to import machinery (like, say, seed processing equipment). International trade is even more difficult and expensive if it has to pass through the destroyed infrastructure of a country like Mozambique (Collier, 2008).

Rich countries that subsidize their agriculture hurt African economies (Collier, 2008). Farm subsidies run at a billion dollars a day, equivalent to the GDP of Africa. Surpluses are dumped on Africa, which lowers the prices African farmers get for their products (Guest, 2004). Foreign companies investing in Africa demand higher returns than on other continents because they perceive risks to be higher (Guest, 2004).

2.4.2 Market institutions

According to Dorward *et al.* (2009) neither state-led nor liberalized market developments have yielded the desired results in Africa. Markets are thin and have high transaction costs. More efforts are needed to better understand and strengthen the institutional environment, including: (i) economic institutions and rules (the political dimension); (ii) values and norms that influence transactions of goods and services; and (iii) social networks.

Marcel Fafchamps argues that, in an unpredictable world, contracts are not always respected. In sub-Saharan Africa, contract agreements are limited by the absence of large hierarchies (both corporate and governmental), so they must depend to a greater degree than in more developed economies on social networks and personal trust (Fafchamps, 2004). Following the privatization in the 1980s, small-scale maize and cowpea seed entrepreneurs in Ghana never established contracts, but, as these businesses grew, they expanded their complex networks of trust with other actors in the system that allowed them access to information and finance (Lyon and Afikorah-Danquah, 1998). Similarly, many of the seed enterprises described in this book have extensive personal ties with the public sector and developed trust with outgrowers, credit suppliers, traders and clients.

Fafchamps (2004) argues that any market transaction is a contract, with mutual obligations and many opportunities for cheating: from faking quality to absconding with payment to diversion of sales by outgrowers to alternative buyers (see also Stringfellow, 1996). For market exchange to take place, buyer and seller must trust each other.

Legal means are not the only way to enforce contracts. Refusing to trade with a person is the most common form of retaliation. Seed dealers who fail to pay their bill to a seed company will soon be marked as non-reliable business people, and depending on the type of interactions between companies may then be refused by the entire seed industry.

Search costs also enforce contracts. If the buyer has gone to a lot of trouble to find a reliable supplier, he or she does not want to start looking all over again, and will be motivated to honour the contract.

In a risky business environment, such as the one in much of the developing world, economic agents face many shocks that make it difficult for them to comply with contractual obligations. African firms may realize that it is difficult for partners to meet all their contractual obligations, and be more lenient with them.

Markets based on 'relational contracting' from social networks can be unfriendly for newcomers, which reduces competition. 'Add a corrupt government and an incestuous banking sector, and you get a business mentality that does not favour growth' (Fafchamps, 2006).

Inadequate market institutions in Africa hinder the activities of small and medium-size firms, which is one reason there are relatively few middle-size firms in sub-Saharan Africa (Fafchamps, 1994, cited in Fafchamps, 2006).

Yet perhaps Fafchamps protests too much and is overly mechanistic. Others argue that in times of uncertainty personal trust becomes even more important. Nigeria has seen its share of ethnic conflict. In 2001, 3000 people died on the Jos Plateau and main markets were destroyed by fire (Porter *et al.*, 2010). (During our fieldwork in 2010 the team was forced to avoid the Jos Plateau because of fighting, and the secretary of one of the seed companies had her house set on fire while she was inside.) Yet Berom tomato retailers in the Jos urban markets can borrow large sums of money from Hausa and Fulani wholesalers without collateral, referring to their long-standing trading relationships. Trust can cut across ethnic lines. People often telephone contacts of other tribes or religions, and tell them not to come to market on a violent day. People say that once in the market they are all Nigerians, that the market belongs to everyone (Porter *et al.*, 2010).

Beer is one of the most commonly sold products in Africa, and is an agro-industrial product. Robert Guest describes a trip to deliver Guinness with a trucker and his assistant in Cameroon. Muddy roads made the 18-hour trip last for 4 days. There were 47 roadblocks and numerous bribes to be paid to the police. Wholesalers in the bush have to carry 5 months' stock (Guest, 2004). Even with these inefficiencies it is profitable to sell beer in Africa, but the brew has certain advantages: fierce demand, low unit price, little specialist knowledge needed to sell or consume it and a long shelf life. Seed has none of those advantages, except low unit price under some conditions.

Molony studied three women in Tanzania exporting high end woodcarvings to the US, Europe and Japan. They all spoke good English. Two had wealthy husbands. They had good contacts locally, started travelling and meeting foreign buyers at trade shows, and now they fund their frequent international travel through the sale of these woodcarvings. They use e-mail to send text messages to clients, but they do not use websites or advertise over the Internet (Molony, 2009).

2.4.3 Financial institutions

A study of traders in Benin and Malawi found that loans and other external finance were rare. Most of the traders owned almost nothing and had few employees. Traders are widely accused of speculating, but the study found that buyers of agricultural commodities did little speculative storage. Most sold as quickly as possible. Buying a bad load could wipe them out, so traders who bought grain almost always transferred it from the sellers' bags to their own bags after buying it, which was tedious but allowed them to assess volume and quality accurately, seeing what was at the bottom of the bag. This quality control meant that the traders had to travel often, to be present at each purchase, adding transaction costs and making it difficult for trading enterprises to grow (Fafchamps and Gabre-Madhin, 2006).

Providing institutional credit for poor borrowers is difficult because: (i) they are usually too poor to provide collateral; (ii) the small loans are too costly to administer individually; and (iii) it is almost impossible to sanction a defaulting borrower. So traditional commercial banks have shown no interest in serving poor farmers. On the other hand, the vast majority of the so-called 'agricultural development banks', which provide subsidized credit, have failed to achieve their objectives both to serve the rural poor and to be sustainable credit institutions (Adams and Vogel, 1985; Braverman and Guasch, 1986). Across sub-Saharan Africa, however, subsidized debt is slowly being replaced by commercial funding, with more emphasis on savings (e.g. saving and credit cooperatives – SACCOs) and more demand-led financial products and services (Green *et al.*, 2006; Nagarajan and Meyer, 2006; AMT, 2008). Farmer seed enterprises, and traders in particular, will benefit from this evolution since they need loans with different payback conditions from agricultural producers. No trader can sell seed at harvest, but must wait until the next cropping season.

Since 1990, group lending with joint liability is the lending model of choice for micro-finance institutions, largely due to the success of the Grameen Bank, which has used the model to achieve high repayment rates on loans to poor rural people of Bangladesh. It is widely believed that the incentives of peer selection, peer monitoring and peer pressure resulting from the joint liability clause are responsible for the high repayment rates. But ample empirical evidence suggests that high repayment rates are also influenced by: (i) the borrower's expectations to access future credit, which would be jeopardized by defaulting on a loan; (ii) screening and monitoring by bank workers in the field; (iii) the local institutional and cultural context; and (iv) the macroeconomic and pricing policies of the government. The failure to appreciate these contributing factors may explain the mixed performance of group lending programmes in other countries (Desai, 1983; Wenner, 1995; Diagne and Zeller, 2001). And even in the case of the Grameen Bank the joint liability clause is often not enforced anyway (Jain, 1996). For seed, even in the informal sector, peer pressure, expectations of future business deals, screening, monitoring and local institutions and culture may also help to ensure that contracts are honoured.

2.4.4 Emerging seed companies

Since the late 1990s, sub-Saharan Africa has seen a four- to fivefold increase in the number of seed companies selling improved maize seed. A survey in Angola, Ethiopia, Kenya, Malawi, Mozambique, Tanzania, Uganda, Zambia and Zimbabwe found that seed companies complained about the high cost of starting a seed company, the difficulties of finding and retaining good staff and of getting source seed, the remoteness of customers and the lack of credit. The companies perceived adoption rates to be low. In some countries where top government officials own shares in seed companies, such companies may be favoured at the expense of others. Most of the enterprises surveyed said that they wanted to produce their own seed, but did not have enough land, so they contracted to outgrowers, although this exposed them to problems from pollen contamination (Langyintuo *et al.*, 2008).

A review published in 2001 found that in Africa, except for a few countries like South Africa and a few crops like hybrid maize, 'there is very little evidence of successful commercial seed sector development' (Tripp and Rohrbach, 2001). In the following chapters we will see that things have changed. Perhaps deregulation and market liberalization have made a difference, along with donor support to market and financial institutions. Before the 1990s, most African countries had large, parastatal seed companies, which failed after structural adjustment. All the parastatals were money losers, producing a limited number of varieties at a high cost (Tripp and Rohrbach, 2001).

The same authors also noted that there was often no budget for maintaining breeder seed, and breeders are rewarded for new varieties, not for maintaining old ones. This led to problems with supply and quality of breeder seed. That much has stayed the same.

Of course, there was political value in handing out free seed, which encouraged governments, donors and NGOs to do so. But the disincentives of these giveaways, the inefficiencies of public seed producers and 'restrictive regulatory regimes' all inhibited the development of a commercial seed sector in Africa (Tripp and Rohrbach, 2001).

2.5 Organizing Farmers to Produce Seed

2.5.1 Neglect by the formal sector

The formal seed sector by itself is generally insufficiently equipped to provide seed for all the smallholder farmers who need it. It is often unprofitable for seed companies to distribute seed to smallholder farmers in low potential areas (Langyintuo *et al.*, 2008), in part because most varieties are bred for high potential areas, but also because reaching many remote smallholder customers requires working with small-scale seed dealers, who often need credit to buy seed and may not be able to store it properly, so they may return unsold seed to the supplier in poor condition.

As a result, private seed companies are often unwilling to deal with many of the improved varieties coming from public research for marginal ecologies. The high costs of producing and distributing certified seed mean that even if such varieties are offered for sale to poor farmers they will only buy a little to try out, and if they like it they will prefer producing their own seed rather than buying it every year.

Improved varieties for marginal conditions may move sluggishly even through informal networks if poor farmers do not produce enough surplus seeds to exchange (Sperling *et al.*, 1996).

2.5.2 Farmers in the formal sector

After disappointing experiences with formal-sector seed projects, 'the idea gained ground that farmers can produce and sell quality seed more cost-effectively than the formal sector'. Basically, the seed provision strategy foresaw that a few specialized farmers or groups could produce seed for a wider area (Almekinders *et al.*, 2007: 367). Thus institutions could sell foundation seed to groups of farmers who would then produce, process and market the seed (Lanteri and Quagliotti, 1997).

A key constraint is the limited market demand for the products of each farmer seed enterprise. To some extent this is because a single village does not provide enough demand to sustain a viable seed business, while farmer seed enterprises usually have few contacts outside their own villages. 'None of these projects have developed retail trading networks' (Tripp and Rohrbach, 2001). And smallholder farmers are reluctant to spend money on quality seed unless it is of a promising new variety (Almekinders and Thiele, 2003). This limits the prices so the farmer seed enterprises may be unable to recover their full costs, and fail (Tripp and Rohrbach, 2001). Most efforts to establish farmer seed-producer groups have had little impact, despite success during the pilot phase.

David (2004) studied three farmer seed enterprises in Uganda. The producers multiplied two bean varieties that had been released in 1994, the first year of her study. They were encouraged to multiply seed of local varieties but showed little interest in doing so, partly because they anticipated low demand. A follow-up survey in 2001 revealed that most of their customers had been one-time buyers. One of the farmer seed enterprises had sold seed to just 4% of the households surveyed in nearby villages. So it sold seed to the district farmers' association and diversified into cassava planting material. The productivity of all three of the enterprises was 'disappointingly low', with yields being 'modest for sole cropping' due to drought, hailstorms, heavy rains, pests and diseases, poor land preparation, late planting, wide spacing, lack of access to land or oxen, and poor soils.

Farmer seed enterprises may be hampered by the lack of adequate resources, as well as facilities such as a well-maintained threshing floor and adequate storage (Tripp and Pal, 2001). One group in Kenya was able to produce high quality maize seed as long as they were given free source seed, fertilizer and pesticides, but their true production costs were well above the sale price of the seed (Chivatsi *et al.*, 2002).

Farmer seed enterprises have been tried scores of times for 15 years and there are basically no success stories. They have problems getting source seed and the donor or NGO does that for them. When the NGO goes, the farmers have nowhere else to turn. They can't do all the storage, inventory, wholesaling and promotion (Tripp, 2003). When projects leave seed sales to farmers, they are rarely able to sell much seed (Tripp, 2001). Many seed projects bypass local seed markets and direct seed company attention away from farmers' needs towards those of projects and governments (Tripp, 2001).

Farmers are good at growing seed, but to make a successful enterprise they may need to link up with other actors, even with input dealers (Tripp and Pal, 2001; Tripp and Rohrbach, 2001; Rubyogo *et al.*, 2010), or else sell to NGOs or government programmes.

Linking multiple actors proved crucial to the success of farmer seed enterprises. In Benin, for about a decade the NGO Songhaï has been marketing its alumni (young agricultural entrepreneurs) by projecting their technical skills and entrepreneurial competences before credit establishments (Dalohoun *et al.*, 2009). Currently, the Banque Régionale de Solidarité (BRS) finances Nerica-related business investments that Songhaï alumni propose for seed and paddy.

2.6 Conclusion

Most efforts to encourage farmers to grow formal seed have failed, and perhaps only succeed where an institution provides transport, technical advice, moral support and marketing; so private companies may be a viable alternative. Of course, it is difficult to maintain a profitable small business anywhere, especially in Africa, and seed is an especially challenging product to sell.

Farmers have reasons for buying seed, even if the reasons are not always as compelling in every case as the formal sector would like to believe. A pattern is emerging for many of Africa's food security crops, especially rice, maize, millet, cowpea and others, where farmers acquire new high yielding varieties and then reproduce the seed on their own farms. This may seem like bad news to private seed companies, but changes in marketing, especially the use of media and small seed packages, may lead to a more regular demand and more stable seed markets.

The examples presented in this book show how various people have made seed enterprises work, either with or without modest external interventions. We hope that these cases will inspire others and reveal some of the underlying keys to success.

References

Adams, D.W. and Vogel, R.C. (1985) Rural financial markets in low income countries: recent controversies and lessons. *World Development* 14(4), 477–487.

Almekinders, C.J.M. and Louwaars, N.P. (1999) *Farmers' Seed Production: New Approaches and Practices.* Intermediate Technology Publications, London. 291 pp.

Almekinders, C.J.M. and Louwaars, N.P. (2002) The importance of the farmers' seed systems in a functional national seed sector. *Journal of New Seeds* 4(1/2), 15–33.

Almekinders, C.J.M. and Thiele, G. (2003) What to do with the seed for small-scale farmers after all? *Cultivos Tropicales* (English version) 24(4), 5–8.

Almekinders, C.J.M., Louwaars, N.P. and de Bruijn, G.H. (1994) Local seed systems and their importance for an improved seed supply in developing countries. *Euphytica* 78(3), 207–216.

Almekinders, C.J.M., Thiele, G. and Danial, D.L. (2007) Can cultivars from participatory plant breeding improve seed provision to small-scale farmers? *Euphytica* 153(3), 363–372.

AMT (African Microfinance Transparency) (2008) *Transversal Analysis of MFI Performance in Africa.* Second Edition, September 2008. AMT, Luxembourg. http://www.microfinance forum.org/cm_data/African_Microfinance_Transparency_–_Transversal_Analysis_2008.pdf. Last accessed 18 April 2010.

Badstue, L.B. (2006) Smallholder seed practices: maize seed management in the central valleys of Oaxaca, Mexico. Thesis, Wageningen University. 327 pp.

Belsky, J.M. and Siebert, S.F. (2004) Cultivating cacao: implications of sun-grown cacao on local food security and environmental sustainability. *Agriculture and Human Values* 20(3), 277–285.

Bentley, J.W. and Vasques, D. (1998) *The Seed Potato System in Bolivia: Organisational Growth and Missing Links.* Agricultural Research and Extension Network (AgREN) Network Paper No. 85. ODI, London.

Bentley, J.W., Tripp, R. and de la Flor, R.D. (2001) Liberalization of Peru's formal seed sector. *Agriculture and Human Values* 18(3), 319–331.

Braverman, A. and Guasch, J.L. (1986) Rural credit markets and institutions in developing countries: lessons for policy analysis from practice and modern theory. *World Development* 14(10/11), 1253–1267.

Brush, S.B. (2004) *Farmers' Bounty: Locating Crop Diversity in the Contemporary World.* Yale University Press, New Haven. 327 pp.

Chivatsi, W.S., Kamau, G.M., Wekesa, E.N., Diallo, A.O. and De Groote, H. (2002) Community-based maize seed production in coastal lowland Kenya. In: Friesen, D.K. and Palmer, A.F.E. (eds) *Integrated Approaches to Higher Maize Productivity in the New Millennium. Proceedings of the 7th Eastern and Southern Africa Regional Maize Conference, Nairobi, Kenya.* CIMMYT, Mexico, D.F., pp. 446–451.

Collier, P. (2008) *The Bottom Billion: Why the Poorest Countries are Failing and What can be Done About It.* Oxford University Press, Oxford. 209 pp.

Dalohoun, D.N., Hall, A. and Van Mele, P. (2009) Entrepreneurship as driver of a 'self-organizing system of innovation': the case of NERICA in Benin. *International Journal of Technology Management and Sustainable Development* 8(2), 87–101.

Daniel, I.O. and Adetumbi, J.A. (2004) Seed supply system for vegetable production at smallholder farms in south-western Nigeria. *Euphytica* 140(3), 189–196.

Daniel, I.O. and Adetumbi, J.A. (2006) Maize seed supply systems and implications for seed sector development in southwestern Nigeria. *Journal of Sustainable Agriculture* 28(2), 25–40.

David, S. (2004) Farmer seed enterprises: a sustainable approach to seed delivery? *Agriculture and Human Values* 21(4), 387–397.

Desai, B.M. (1983) Group lending in rural areas. In: von Pischke, J.D., Adams, D.W. and Gordon, D. (eds) *Rural Financial Markets in Developing Countries: Their Use and Abuse.* Johns Hopkins University Press, Baltimore, MD, pp. 284–288.

De Schutter, O. (2009) *The Right to Food Seed Policies and the Right to Food: Enhancing Agrobiodiversity and Encouraging Innovation.* Report to the UN General Assembly, New York. 22 pp.

Diagne, A. and Zeller, M. (2001) Access to Credit and its Impact on Welfare in Malawi. *IFPRI Research Monograph* No. 116. IFPRI, Washington.

Dorward, A.R., Kirsten, J.F., Omamo, S.W., Poulton, C. and Vink, N. (2009) Institutions and the agricultural development challenge in Africa. In: Kirsten, J.F., Dorward, A.R., Poulton, C. and Vink, N. (eds) *Institutional Economics Perspectives on African Agricultural Development.* IFPRI, Washington, pp. 3–34.

Doss, C., Mwangi, W., Verkuijl, H. and de Groote, H. (2003) *Adoption of Maize and Wheat Technologies in Eastern Africa: a Synthesis of the Findings of 22 Case Studies.* CIMMYT Economics Working Paper 03–01. CIMMYT, Mexico, D.F. 32 pp.

Easterly, W. (2006) *The White Man's Burden. Why the West's Efforts to Aid the Rest have Done So Much Ill and So Little Good.* Penguin Books, London. 436 pp.

Fafchamps, M. (1994) Industrial structure and microenterprises in Africa. *Journal of Developing Areas* 29(1), 1–30.

Fafchamps, M. (2004) *Market Institutions in Sub-Saharan Africa.* MIT Press, Cambridge, Massachusetts. 521 pp.

Fafchamps, M. (2006) *Spontaneous Markets, Networks, and Social Capital: Lessons from Africa.* Working paper, series 058. Global Poverty Research Group, London.

Fafchamps, M. and Gabre-Madhin, E. (2006) *Agricultural Markets in Benin and Malawi.* Working paper, series 068. Global Poverty Research Group, London.

Gabre-Madhin, E.Z. and Haggblade, S. (2004) Successes in African agriculture: results of an expert survey. *World Development* 32(5), 745–766.

Gildemacher, P.R., Demo, P., Barker, I., Kaguongo, W., Woldegiorgis, G., Wagoire, W.W., Wakahiu, M., Leeuwis, C. and Struik, P.C. (2009) A description of seed potato systems in Kenya, Uganda and Ethiopia. *American Journal of Potato Research* 86(5), 373–382.

Green, C.J., Kirkpatrick, C.H. and Murinde, V. (2006) Finance for small enterprise growth and poverty reduction in developing countries. *Journal of International Development* 18(7), 1017–1030.

Guest, R. (2004) *The Shackled Continent: Africa's Past, Present and Future.* Macmillan, London. 280 pp.

Jain, P.S. (1996) Managing credit for the rural poor: lessons from the Grameen Bank. *World Development* 24(1), 79–89.

Jones, R.B., Audi, P.A. and Tripp, R. (2001) The role of informal seed systems in disseminating modern varieties. The example of pigeonpea from a semi-arid area of Kenya. *Experimental Agriculture* 37(4), 539–548.

Kloppenburg, J.R., Jr (1988). *First the Seed: the Political Economy of Plant Biotechnology, 1492–2000.* Cambridge University Press, Cambridge. 425 pp.

Langyintuo, A.S., Mwangi, W., Diallo, A.O., MacRobert, J., Dixon, J. and Bänziger, M. (2008) *An Analysis of the Bottlenecks Affecting the Production and Deployment of Maize Seed in Eastern and Southern Africa.* CIMMYT, Harare, Zimbabwe.

Lanteri, S. and Quagliotti, L. (1997) Problems related to seed production in the African region. *Euphytica* 96(1), 173–183.

Lipper, L., Anderson, C.L. and Dalton, T.J. (eds) (2009) *Seed Trade in Rural Markets: Implications for Crop Diversity and Agricultural Development.* FAO, Rome. 224 pp.

Longley, C., Jones, R., Ahmed, M.H. and Audi, P. (2001) *Supporting Local Seed Systems in Southern Somalia: a Developmental Approach to Agricultural Rehabilitation in Emergency*

Situations. Agricultural Research and Extension Network (AGREN) Network Paper No. 115. ODI, London.
Lyon, F. and Afikorah-Danquah, S. (1998) *Small-scale Seed Provision in Ghana: Social Relations, Contracts and Institutions for Micro-enterprise Development*. Agricultural Research and Extension Network (AGREN) Network Paper No. 84. ODI, London.
Marfo, K.A., Dorward, P.T., Craufurd, P.Q., Ansere-Bioh, F., Haleegoah, J. and Bam, R. (2008) Identifying seed uptake pathways: the spread of Agya Amoah rice cultivar in southwestern Ghana. *Experimental Agriculture* 44(2), 257–269.
Meredith, M. (2006) *The State of Arica: a History of Fifty Years of Independence*. Free Press, London. 752 pp.
Molony, T. (2009) Carving a niche: ICT, social capital, and trust in the shift from personal to impersonal trading in Tanzania. *Information Technology for Development* 15(4), 283–301.
Nagarajan, G. and Meyer, R.L. (2006) Rural finance today: advances and challenges. *Finance for the Poor* 7(4), 1–8.
Porter, G., Lyon, F., Adamu, F. and Obafemi, L. (2010) Conflict and cooperation in market spaces: learning from the operation of local networks of civic engagement in African market trade. *Human Organization* 69(1), 31–42.
Poulton, C., Kydd, J. and Dorward, A. (2006) Overcoming market constraints on pro-poor agricultural growth in sub-Saharan Africa. *Development Policy Review* 24(3), 243–277.
Rohrbach, D.D., Minde, I.J. and Howard, J. (2003) Looking beyond national boundaries: regional harmonization of seed policies, laws and regulations. *Food Policy* 28(4), 317–333.
Rubyogo, J.C., Sperling, L., Muthoni, R. and Buruchara, R. (2010) Bean seed delivery for small farmers in sub-Saharan Africa: the power of partnerships. *Society and Natural Resources* 23(4), 285–302.
Sperling, L. (2002) Emergency seed aid in Kenya: some case study insights on lessons learned during the 1990s. *Disasters* 26(4), 329–342.
Sperling, L. and Cooper, H.D. (2003) Understanding seed systems and strengthening seed security: a background paper. In: *Improving the Effectiveness and Sustainability of Seed Relief: a Stakeholders' Workshop*. Food and Agriculture Organization, Rome.
Sperling, L., Scheidegger, U. and Buruchara, R. (1996) *Designing Seed Systems with Small Farmers: Principles Derived from Bean Research in the Great Lakes Region of Africa*. Agricultural Research and Extension Network (AGREN) Network Paper No 60. ODI, London.
Stringfellow, R. (1996) *Smallholder Outgrower Schemes in Zambia*. ODA Crops Post-harvest Programme, Project No. AO439. Natural Resources Institute, Chatham. 37 pp.
Thiele, G. (1999) Informal potato seed systems in the Andes: why are they important and what should we do with them? *World Development* 27(1), 83–99.
Tripp, R. (2001) *Seed Provision and Agricultural Development: the Institutions of Rural Change*. Overseas Development Institute, London. 174 pp.
Tripp, R. (2003) How to cultivate a commercial seed sector. Sustainable agriculture in the Sahel. Paper prepared for the symposium 'Sustainable Agriculture in the Sahel', Bamako, Mali, 1–5 December 2003.
Tripp, R. and Pal, S. (2001) The private delivery of public crop varieties: rice in Andhra Pradesh. *World Development* 29(1), 103–117.
Tripp, R. and Rohrbach, D.D. (2001) Policies for African seed enterprise development. *Food Policy* 26(2), 147–161.
Tripp, R., Louwaars, N. and Eaton, D. (2007) Plant variety protection in developing countries. A report from the field. *Food* Policy 32(3), 354–371.
Wenner, M.D. (1995) Group credit: a means to improve information transfer and loan repayment performance. *Journal of Development Studies* 32(2), 263–281.

3 Cameroon: Revolving Funds Make a Difference

DRISSA SILUÉ, ABDOULAYE BARRA
AND ROBERT G. GUÉI

3.1 Introduction

3.1.1 Agriculture

Cameroon has 18 million inhabitants, 62% rural; 1.1 million farms have less than 2 hectares, although there are some large government and private farms. Most farms (72%) manage both crops and livestock; 25% are specialized in crops only (INS, 1999). Agriculture contributes 20% of the gross domestic product (FAR, 2010) and employs 73% of the workforce (Bella, 2009).

Of the 9.2 million hectares of potential agricultural land, only 1.8 million are farmed (FAR, 2010). Commercial farms produce export crops like cocoa, cotton, rubber, sugarcane, banana, oil palm, tea and tobacco. These large government and private companies play a big role in the economy. Smallholder farms produce staple food with 72% growing maize or sorghum, 71% groundnut, 58% root crops like taro or coco yam, 56% plantain, 53% bean or cowpea and 52% vegetables. Many also grow cassava, sweet potato or potato, among many other crops. Nearly 70% of the food produced is eaten on-farm (Annuaire Statistique du Cameroun). Rice is grown by an estimated 145,000 farmers on about 44,000 hectares, mainly in the irrigated schemes in the north. As these are far removed from the centres of consumption in the south (Yaoundé and Douala), most rice is exported to the neighbouring countries of Nigeria, Chad and the Central African Republic (MARD, 2009).

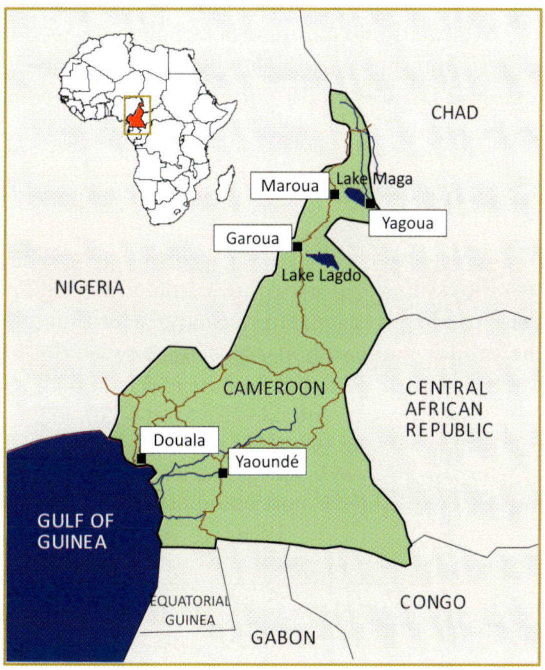

Most of Cameroon has two rainy seasons, although the north has only one (June to November), with rains being more erratic in the far north.

Because of the low use of farm inputs and low soil fertility, the yields of most staple food crops are low in Cameroon, 2 tonnes per hectare for maize, 0.8 for sorghum, 0.8 for groundnut, 12 for cassava and 14 for plantain.

Food production is lower than demand, with imports making up the difference. In Central Africa, rice imports increased 14-fold between 1961 and 2007, rising from 32,100 to 470,974 tonnes, while per capita cereal production shrank from 157 to 85 kg (MARD, 2009). This is also true for Cameroon, which imported 300,000 tonnes of rice in 2009 while only producing 100,000 tonnes of paddy.

Although major roads are good in northern and southern Cameroon, the regions are poorly linked, apart from air travel. With increasingly erratic rainfall affecting agriculture in the drier north and no possibility of transporting food from the major port in Douala, FAO decided to help northern and far northern Cameroon become food secure. The seed enterprises described in this chapter focus on these two zones.

3.1.2 Seed industry

There are no private seed companies producing or selling seed in northern and far northern Cameroon. State seed agencies have stopped producing seed and their land is now used by smallholder farmers either free of charge or for rent. Currently farmer seed producer groups are the only ones producing and selling seed.

Multinational companies like Pioneer Seed tried to produce seed on government-owned land, but failed. The official reason given was that seed production costs were high and farmers could not afford to buy seed at the proposed prices.

3.1.3 Seed legislation

In 2001, a law was issued on seed prices stipulating that seed must be produced under specific terms of references to earn certain high prices. As no seed producer is meeting these quality standards, prices are determined by the market. Seed prices depend on the vendor and location and can be twice the seed producer's price. For example, in 2009 rice seed was sold at 200 FCFA per kg ($0.48) in Lagdo area, while in Maga and Yagoua it was sold at 400 FCFA ($0.95).

A law signed in 2005 created a fund to support the seed industry and seed research and to improve the conservation of farm-saved seed. A second law signed in 2005 regulates seed certification, quality control and marketing.

The law recognizes two types of seed: certified and local, with no intermediate category. Certified seed production in Cameroon has been around 7000 tonnes (Table 3.1).

Certification starts by declaring one's intent to produce seed before planting, followed by at least three field visits, sampling at harvest and issuing of certificates. However, laboratory analyses are still limited to seed germination and purity, as there is no equipment yet to test for pests and diseases.

So, by law, seed sold has to be certified, but, as the demand for seed far exceeds supply and the certification system is not yet fully operational, there is no enforcement of the regulation and any seed is accepted for sale. Fraud control is not yet implemented, although the Ministry of Agriculture and Rural Development is planning to strengthen seed inspection and certification. FAO recently trained inspectors and will contribute to making seed laboratories available throughout the country.

Table 3.1. Seed certified (tonnes) in Cameroon, 2006–2009.

	2006	2007	2008	2009
Maize	3,457	4,075	4,100	4,124
Rice	575	625	710	718
Sorghum	81	125	120	125
Millet	10	14	11	14
Potato	1,551	1,923	1,810	1,940
Cowpea	110	195	213	217
Groundnut	76	100	91	94
Beans	36	48	50	48
Soybean	9	11	11	13
Total	5,903	7,116	7,106	7,293

3.1.4 Creating a sustainable seed production and supply system

When state agencies stopped producing seed there were no alternative sources. To address food insecurity in northern and far northern Cameroon the FAO and the government of Cameroon started two projects, Support for the multiplication and diffusion of improved and healthy rice seeds (Appui à la multiplication et à la diffusion de semences améliorées et saines de riz) (2004 to 2006) and Support to farmer organizations for the multiplication and diffusion of early maturing varieties of maize, sorghum and millet in the provinces of the north and far north (Appui aux organisations paysannes pour la multiplication et la diffusion de variétés précoces de maïs, de sorgho et de mil dans les provinces du Nord et de l'Extrême Nord) (2006 to 2008).

Both regions are prone to drought and needed support in grain production. State facilities for irrigation (dams and water reservoirs) were available. The governmental development authority MEADEN (Mission d'Études pour l'Aménagement et le Développement de la province du Nord) looks after the management of the Lagdo dam area, 60 km from Garoua, the regional capital in the northern region. Another governmental development institute, the authority for the expansion and modernization of rice in Yagoua (SEMRY – Société d'Expansion et de Modernisation de la Riziculture de Yagoua), completed a dam in Maga on the Logone River in 1979, specifically to irrigate rice in the far northern region. The Institute of Agricultural Research for Development (IRAD) had developed drought tolerant and short cycle crop varieties of maize, sorghum and millet and seed was available, albeit not in farm communities.

Triggered by the two FAO projects, certified seed is currently produced and sold in labelled bags in northern and far northern Cameroon (Table 3.2). Rice seed produced from 2005 to 2009 could not be officially certified as the foundation seed used came from the governmental development institute SEMRY, whereas by law this must come from IRAD. The rice seed produced was therefore not recorded or included in this table.

Table 3.2. Certified seed produced (tonnes) under FAO project.

	2007	2008	2009
Maize	230	202	207
Sorghum	41	18	16
Pearl millet	4	3	0
Total	275	223	223

The two FAO projects aimed to strengthen cereal seed production and dissemination by farmer seed enterprises. Seed producer groups were formed or strengthened and officially registered as legal entities (GIC – groupement d'initiatives communes) and established their own revolving funds for seed production and marketing.

Common features of the farmer seed enterprises. Many of the enterprises created during the FAO projects have so far sustained their activities after the projects ended. They all set up and are managing their own revolving funds.

The seed produced was meant for local farmers of each project area, but some was sold to Chad and Niger. Before the project, paddy yields obtained with local seed in irrigated plots were below 4 tonnes per hectare while yields with quality seed were at least 8 tonnes after the project.

Evolution of the seed producer groups. In 2006, by the end of the 2-year rice project, 25 groups produced rice seed. In 2007, this number had dropped to eight, but seven groups took up seed production again in subsequent years (Table 3.3).

During the second project 117 sorghum, maize and millet groups registered in 2007, of which by 2008 111 were still active. This further dropped to 68 groups a year after the project ended (58% of the initial number) (Table 3.4).

For all cereal crops, 83 groups were still operating in northern and far northern Cameroon in 2009. All the groups said that demand for seed is still greater than what they can supply and they plan to expand their enterprises.

The variation in group numbers can be attributed to three factors: (i) some groups became landless as the landlord decided to stop renting them land and to embark on seed production himself as the activity was becoming profitable; (ii) some groups were

Table 3.3. Evolution of rice seed producer groups.

Region	2005	2006*	2007	2008	2009
North					
Number of groups created	0	4	0	0	0
Number of groups which disappeared	0	0	1	0	0
Number of groups active	0	4	3	3	3
Far North					
Number of groups created	8	13	0	4*	3*
Number of groups which disappeared	0	0	16	0	0
Number of groups active	8	21	5	9	12
Total number of active groups	8	25	8	12	15

*Immediately after the FAO project ended in 2006 some groups stopped producing seed, but restarted a few years later.

unable to properly manage their revolving fund scheme; and (iii) some groups merged to form larger groups.

Table 3.4. Evolution of maize, sorghum and millet seed producer groups.

Region	2007	2008*	2009
North			
Number of groups created	32	9	0
Number of groups which disappeared	0	6	3
Number of groups active	32	35	32
Far North			
Number of groups created	85	0	0
Number of groups which disappeared	0	9	40
Number of groups active	85	76	36
Total number of active groups	117	111	68

*End of FAO project.

Structure. All newly established enterprises are officially registered and organized in a similar way with a president, a secretary, a treasurer and an accountant. The group members conduct all activities together, including seed production (ploughing, nursery, planting, harvesting, seed treatment and packing) and marketing (sales in markets or to farmers). Basically, the group members are the enterprises' staff. However, during labour-intensive activities such as ploughing or harvesting, the groups hire labourers. Specific details for each enterprise are described below.

Links. The project provided all enterprises with functional linkages with either MEADEN or SEMRY, both government development authorities. MEADEN manages the Lagdo dam area in the north. It cleans irrigation canals, manages the water supplies (for fees) and gives free technical advice to rice producers. MEADEN grants land for seed production to selected groups free of charge. However, farmers are responsible for buying inputs and for ploughing their plots. Being on good terms with MEADEN is important, as seed cannot be produced without land.

SEMRY aims at improving and expanding rice production in the Maga dam area, around Yagoua in the far northern region. It provides ploughing services, irrigation water, cleaning of the irrigation channels and advisory services. Land is owned by SEMRY, which grants each producer or group plots for a fee of 51,000 FCFA per hectare ($121).

Capacity. Building technical and managerial capacities of the groups is essential for sustainability. The farmer groups received training in seed production, processing, packaging and marketing. They also learned how to manage a farmer seed enterprise and how to set up and manage a revolving fund. They received quality seed, fertilizers, pesticides and bags for seed packing. MEADEN and SEMRY published leaflets on seed production and distributed them to all groups. All groups became competent to produce and market their seeds, although some refresher courses are needed.

Setting up the revolving funds. Having easy access to formal financial services is often impossible in rural areas and constitutes a major bottleneck to the growth and development of seed enterprises. The FAO projects helped the groups open bank accounts at Credit du Sahel, which was created in 1997 by individuals in northern Cameroon and which aims at reducing poverty. It has 17 branches and 13 village banks in the three northern provinces of Cameroon.

The FAO projects advanced money to the seed producer groups to plant and market their first seed crop. After selling the seed from the first harvest the groups deposited the amount advanced into their own bank accounts, and added another 15% of the seed sales to it. This money was then available for the next planting season or served to secure loans from the bank.

The projects also subcontracted specialized institutions like IRAD and consultants to train the groups in managing this revolving fund. All training was provided at no cost. To increase their fund, groups decided that each member would donate a certain number of rice bags. All bags are then sold and the collected funds are used to replenish the revolving funds. The funds are first of all used to fund collective farm activities and have been essential in sustaining group activities after the projects ended.

Seed inspection and certification. Ensuring the quality of seeds produced is important to build trust, to guarantee a market and obtain a good price. In the FAO project areas the seed inspection service (Service Régional de Contrôle des Intrants et des Produits Agricoles), based in Garoua, has two inspectors and one lab technician. The laboratory is under renovation and is expecting new equipment. Each farmer of a group planning to produce seed has to declare the location of his or her plot, its size and the crop. Currently the staff makes three inspections (at planting, before flowering and at harvest). Seed samples are analysed in the laboratory for purity and germination. Labels and packaging are only for seed with more than 87% germination. The inspection and certification cost is supported by the government through the Seed Fund. Soon the inspectors also intend to inspect markets, looking for fraudulent seeds.

3.2 The Aoudi Sanguéré Federation

3.2.1 History

The Aoudi Sanguéré Federation was created in 1994 and hosts 14 unions with 2000 members. It is based in Garoua, is officially registered and operates at the national level. The federation was created by former MEADEN and SEMRY employees who knew seed production. Because of its capacity to produce foundation seed, the federation was included in both FAO projects.

The federation obtains breeder seed from IRAD and produces foundation and quality seed of maize, rice, sorghum, cowpea and groundnut, the main crops in northern and far northern Cameroon. Since 2005 the federation has produced 57 tonnes of foundation seed, of which 80% was maize (Table 3.5).

The seed is produced on its own land during the rainy season (June to December) only because irrigation is not available. Seed production has increased over the years and amounted to 30 million FCFA ($71,400) in 2009. Some of the federation members also produce quality seed on their own land.

The federation intends to increase its membership to 3000 by accepting selected new groups and individual farmers willing to join. The federation also plans to evaluate some Nerica rice varieties in the region and set up a commission for investigating fraudulent seed sales. It will work for stronger law enforcement to avoid grain being sold as seed.

3.2.2 Structure

Table 3.5. Foundation seed produced (tonnes), Aoudi Sanguéré Federation.

	2005	2006	2007	2008	2009
Maize	9.2	8.3	6.4	13.0	9.0
Rice	0.7	0.3	0	0.5	0.6
Cowpea	1.4	3.0	0	0.7	0.2
Groundnut	1.1	1.3	0.3	0.8	0.3
Total	12.4	12.9	6.7	15.0	10.1

The federation has a management committee composed of a secretary general, a treasurer, an internal auditor, a storekeeper and a technical supervisor. It has noticed that some of the varieties in circulation are not pure and give low yields. Besides purifying them, it explores sources of pure breeder and foundation seed of the varieties that it grows. They work in an office in Garoua, along with several casual labourers. The federation plans to increase its farm size, which in 2009 was about 7 hectares.

They apply fertilizers, herbicides and pesticides, but no irrigation. They use a big warehouse belonging to the Regional Agriculture and Rural Development authority which can accommodate 100 tonnes of seed. The federation is responsible for maintaining the warehouse. Recently, it has asked the local authority for a 1000 square metre plot on which it wants to build its headquarters, a seed warehouse, a seed drying facility and a small seed quality control laboratory for internal usage. Individual members also grow vegetables and rear small animals, such as goats, chickens and sheep.

3.2.3 Cash flow

The federation has a revolving fund of 2,000,000 FCFA ($4760), successfully implements farm activities and markets its seeds without any external support. It is planning to increase its revolving fund in 2010 by raising the contribution of individual member farmers and increasing its membership.

Among other things, the revolving fund is used to pay a local supplier of seed bags with whom the group has a long-standing relation. The cost of the bags is recovered through seed pricing to enable continuous sourcing of the bags, which hold between 2 and 15 kg depending on the crop.

3.2.4 Marketing

Main clients for the foundation and certified seed are the 2000 members of the 14 unions, who are all seed producers. Other clients include local seed and grain producers. In 2009, certified seed was also sold to relief agencies for Chad (Table 3.6).

To boost its seed sales, the federation organizes radio campaigns and participates in seed fairs and agricultural shows. The federation also organizes open days to demonstrate quality and transparency.

The federation will also sign partnership agreements and engage in the distribution of seed produced by these partners to further increase the income of the group.

3.3 The Sayem Seed Producers Union

3.3.1 History

The Sayem Seed Producers Union (Union de groupements de multiplicateurs de semences de Maga) comprises three rice producer groups with 12 to 16 members. The groups were created in 2005 with FAO support and are legally registered. The union is based in the Maga dam area, 300 km from Garoua, in the far northern region.

The governmental development authority SEMRY manages 11,500 hectares with 6200 allocated to rice producers at Maga, although the area has a potential of 20,000 hectares of irrigated rice (MARD, 2009). When the irrigated area is increased to 20,000 hectares, the union will be able to increase its production.

Before the FAO project, paddy yields obtained in Maga with local seed were below 4 tonnes per hectare, while they now reach 8 t because of access to quality rice seed.

One rice variety, IR 46, has been grown in the Maga area since the beginning of the project. Two cropping seasons are possible but land is scarce and the union produces seed only in the dry season (December to May). Seed of IR 46 could be grown during the rainy season, but it is highly susceptible to rice blast, a serious disease caused by the fungus *Magnaporthe grisea*.

Farmers would also like to have other varieties, especially ITA 300, which is high yielding and grows in the rainy season. Since the Sayem Union was established, they have produced about 490 tonnes of quality seed (Table 3.7). The union produces more than 70% of

Table 3.6. Clients of the Aoudi Sanguéré Federation.

	1994	2009	2015 (predicted)
Members	1	1	1
Relief agencies	–	2	3
Individual farmers	–	3	2
Women and youth groups	–	–	4
Companies	–	–	5
Agro-dealers	–	–	6

Ranking assessment by senior management of seed enterprise, 1 being the most important.

Farmers in northern Cameroon grow rice seed only in the dry season, because their popular variety, IR 46, is susceptible to rice blast, especially during the rainy season.

Table 3.7. Quality rice seed produced, Sayem Union.

	2005	2006	2007	2008	2009
Tonnes	72	132	109	–*	176
Hectares	15	22	21	–	22

*No plot was granted to the union.

the total seed needed for the Maga region and the group plans to meet the regional need for quality seed.

3.3.2 Structure

Land is the main limitation to producing rice seed in the Maga region, since all 6500 hectares are distributed to farmers who want to produce rice. Land is allocated based on family size, but no farmer can access more than 2 hectares. SEMRY allocates land directly to the union for seed production and its members also apply as individuals to SEMRY for land for paddy or seed production. Since its creation the union has planted 22 hectares.

For both quality seed and paddy production, fertilizers are used but urea is expensive and not always available on time. Pesticides are also used. Machinery for harvesting and threshing is not available. The groups store their own seed until it is sold.

Irrigation permits a second growing season per year, but farmers in the Maga region currently grow rice only during the dry season because the rice varieties they use have a long cycle and, if grown during the rainy season (June to November), occasional rains towards the end of the season make seed harvesting, drying and packing risky. Farmers buy the foundation seed for quality seed production from the Aoudi Sanguéré Federation. Farmers also raise livestock and have other activities.

3.3.3 Cash flow

The revolving fund is big enough to grant interest-free loans to members. The fund is also used to support social activities, such as weddings, funerals, school fees and receiving visitors. Members who benefit from such grants pay them back to the group. For now, seed supply is still below the demand.

3.3.4 Marketing

At the beginning of the project quality seed was sold to FAO. Currently, all seed is sold to local farmers, who are attracted by the quality of the seed and the resulting high yields, as demonstrated on field days. Occasionally, quality seed or paddy is also sold in Niger (Niger Basin Project). The union is looking for big customers, but it is aware that the main limitation will be access to enough land to produce the quantities requested. Seed produced is stored and marketed collectively.

Seed price has been stable since the union started and is currently 400 FCFA per kg ($0.95) while paddy is sold for 150–210 FCFA ($0.36–0.50). The members earn a substantial income from rice seed, but also from paddy (at least 2 hectares per member), fishing, livestock and vegetable growing.

At the beginning of the project, FAO provided bags for packing the seed, holding 2 kg for maize seed and 5 kg for rice seed. Sales increased thanks to the prestigious FAO logo on the bags, promising high quality. This experience has created an incentive for seed producers to embark on branding their products.

3.4 The Agrelenas Seed Producers Group

3.4.1 History

The Agrelenas group (Groupement de producteurs de semences de maïs de Nassarao-Garoua II) was created in 2006 by the FAO and is based in Garoua. This group started with 12 members but now has eight men and one woman.

All varieties used came from IRAD during the FAO project. Three maize varieties are used: CMS 90-15, CMS 85-01 and CMS 87-04, which yield on average 3 tonnes per hectare. The rice variety used is IRAT 112, which yields on average 4.5 tonnes per hectare. The maize and rice seed produced has slowly increased (Table 3.8).

Table 3.8. Quality seed produced, Agrelenas group.

	2007	2008	2009
Maize	2.5 (2)*	2.0 (2)	3.0 (4)
Rice	3.0 (0.5)	4.5 (1)	5.0 (1)

*In tonnes, with hectares cultivated in parentheses.

3.4.2 Structure

The group does all its activities together. Land is rented and is difficult to find near their homes around Garoua. The group has to rent it further and further away due to competition and prices. When some landlords see that seed production is lucrative, they decide to keep the land to start producing seed themselves. The group rents 12 hectares of land at 20,000 FCFA ($48) per hectare. The group has still not been permanently settled on land.

Seed producers buy foundation seed from the Aoudi Sanguéré Federation in Garoua (Section 3.2) to grow quality seed during the rainy season.

Fertilizers are used when they are available on time and affordable. Pesticides are also used. Farm machinery is not available and all work is done by hand. They have a storehouse and are building a bigger one.

Most of the members of this group were involved in maize, rice and off-season sorghum grain production. Income from seed is now 40% of their income. Some of the members produce vegetables, some keep cows, goats, sheep and chickens.

The group is currently investing in a warehouse and has acquired two oxen for ploughing the land. It will significantly increase its revolving funds and agricultural land. It has acquired a water pump for vegetable production as a group.

3.4.3 Cash flow

The group has a revolving fund and is successful without outside support. It has a bank account and access to micro-credit. They have not used bank loans. The revolving fund holds a minimum of 700,000 FCFA ($1670) for production and marketing. Members can also get loans without interest from this fund for funerals, school fees, Ramadan and Christmas.

3.4.4 Marketing

Initially the seed was sold to local farmers, the FAO project and the Ministry of Agriculture, but currently only to local grain producers. The seed is produced during the rainy season (June to December) only.

The price of seed has been stable at 500 FCFA per kg ($1.19) as against 100–200 FCFA per kg for grain ($0.24–$0.48). The group stressed that the seed bags provided by the FAO were good publicity. They plan to produce similar bags.

Most groups use signboards to announce the seed varieties grown and a contact telephone number.

3.5 Challenges and Strengths of the Seed Enterprises

Limited access to land. In the government managed areas (SEMRY and MEADEN) land is ploughed and distributed to groups or individual farmers according to availability and other criteria (such as family size). In Maga two rice seasons are possible and farmers and common initiative groups have to pay 51,000 FCFA ($121) per hectare, but land is still not available to grow rice during the rainy season. For farmers outside the government managed plots, access to land is uncertain, especially near the farmstead (Goufo, 2008). Long distances increase transport costs. In the government strategic plan (MARD, 2009) irrigated land will be increased in the Lagdo dam area to 5000 hectares; it is now 800. In the Maga dam area SEMRY available land will be increased from 11,500 hectares to 20,000. However, corruption in land allocation has often raised tensions and weakened local institutions governing collective natural resource management (Asah *et al.*, 2008).

Availability of farm inputs. All groups complained of high input prices, especially for fertilizers. The price of NPK doubled in 2007, and it was often not available on time.

Availability of machinery. Machinery for ploughing, harvesting and threshing is not available for hire. For ploughing the groups hire ox teams; they hire casual labourers to help with the harvest.

Revolving funds. Revolving funds for seed production were the base for success of all these enterprises. Some groups, not described in this chapter, failed because they did not set up a revolving fund. The successful ones all set up funds of at least 700,000 FCFA ($1670). Such a small fund is used exclusively for production and marketing. Groups with bigger funds make short-term, interest-free loans for school fees, funerals and weddings. All groups are planning to increase their funds after the next harvest.

In general, financial institutions lend more easily to groups rather than to individuals as risks and costs are lower (Adams and Ladman, 1979). To better equip farmers to cope with change in the drought-prone areas of northern Cameroon, autonomous saving and credit systems at the village level need further strengthening (Asah *et al.*, 2008).

Availability of pure foundation seed. Some of the foundation seed does not meet quality standards, leading to low yields. Seed growers need pure seed lots which give higher yields and quality of a level acceptable to the seed inspectors.

Need to better organize the seed industry. The unions want the government to reinforce the seed law in order to prevent any fraudulent sale of low quality seeds or grains. The seed growers intend to set up a commission to improve the collection, storage and marketing of quality seed. Federations can make significant contributions in seed marketing and as such help to sustain their farmer seed producer groups (Samsuzzaman and Van Mele, 2005).

Factors of success. The seed producers said that the key factor to success was group solidarity and motivation followed by the high incomes they were able to achieve. The regular visits of the FAO project manager had initially motivated them and showed them that seed production was important. Most members have made a major purchase (a house, a shop or a motorbike). All members have added new activities such as individual seed production, animal husbandry, grain and vegetable production.

The successful revolving fund and self-financing led to success; all farm operations are now self-funded. All enterprises are planning to increase the fund beyond the current minimum of 700,000 FCFA ($1670).

High seed quality (all the seed produced is inspected by the government) was also important. Most grain producers got higher yields and returned to buy quality seed. Farmers who used the quality seed and got higher yields spread the news to their neighbours, increasing demand and markets.

References

Adams, D.W. and Ladman, J. (1979) Lending to the rural poor through informal groups: a promising financial market innovation. *Saving and Development* 3(2), 85–94.

Asah, S.T., Nelson, K.C. and Bengston, D.N. (2008) Managing social-ecological change and uncertainty: floodplain agriculture and conservation in Dryland northern Cameroon. *Conservation and Society* 6(2), 166–178.

Bella, H. (2009) *Agriculture et croissance économique au Cameroun.* Institut Sous-régional de Statistique et d'Economie Appliquée (ISSEA), Mémoire de fin de formation, Ingénieur d'Application de la Statistique. http://www.memoireonline.com/10/09/2761/Agriculture-et-croissance-economique-au-Cameroun.html.

FAR (Réseau International Formation Agricole et Rurale) (2010) *Cameroun.* http://www.far.agropolis.fr/infospays/cameroun.php.

Goufo, P. (2008) Evaluating the constraints and opportunities for sustainable rice production in Cameroon. *Research Journal of Agriculture and Biological Sciences* 4, 734–744.

INS (1999) *Annuaire Statistique du Cameroun 99. DSCN, novembre 1999.* Institut National de la Statistique, Yaoundé.

MARD (2009) *National Strategy for Rice Growing in Cameroon.* Ministry of Agriculture and Rural Development, Yaoundé. http://www.jica.go.jp/activities/issues/agricul/pdf/NRDS/NRDS_came_E.pdf.

Samsuzzaman, S. and Van Mele, P. (2005) Innovating with federations: community institutions take the lead in seed marketing. In: Van Mele, P., Salahuddin, A. and Magor, N.P. (eds) *Innovations in Rural Extension: Case Studies from Bangladesh.* CABI Publishing, Wallingford, pp. 245–254.

4 Nigeria: Clustered Seed Companies

Jeffery W. Bentley, Olupomi Ajayi and Kehinde Adelugba

4.1 Introduction

4.1.1 Agriculture

Northern Nigeria is flat, with a long dry season. The south is hillier with a longer rainy season. The main crops in the north are maize, sorghum, millet, rice, groundnut and irrigated crops, especially vegetables. The main crops of the south are roots and tuber crops such as yam, cassava, coco yam and sweet potatoes, as well as oil palm, mangoes, citrus, pineapple and other fruits.

Nigerian farms are small, an average of some 2 hectares, and about 45% of the gross national product is from agriculture. The British colonial government did not create large estates, but did establish commodity boards, monopolies which paid fixed prices for groundnuts, cocoa, palm oil and cotton for export. The boards lived on after independence in 1960, but were disbanded during the structural adjustment of the 1980s.

Roads are gradually improving, with divided highways (freeways, motorways) now connecting some major cities, but even some of these freeways are now plagued with potholes, adding to transportation costs for agriculture. The electricity is off for most of the day, forcing many agro-industries to spend precious capital on fuel and generators. Federal and state governments have subsidized fertilizer and seed for some years now, thereby strengthening the demand for seed.

4.1.2 Seed industry

There are 13 companies registered to produce certified seed in Nigeria. There are about nine outgrowers, producing foundation seed for the National Agricultural Seed Council. Farmers' associations produce about 12–20% of the certified seed used in Nigeria. They exist in almost every state. The first seed company was started in 1984 (Section 4.4.1) and, although it is a difficult market, new companies, often small ones, have started constantly since then. NGOs play less of a role in Nigeria than in some other African countries. NGOs produce little certified seed (Table 4.1), and only one seed company mentioned them as important customers. There are only two farmers' associations registered to produce seed in Nigeria.

What is unusual about Nigeria is the large presence of local government in seed production and distribution. Nigeria is divided into 36 states plus the Federal Capital Territory, roughly like the US model. Many of the Nigerian states have seed production units in their ADPs (agricultural development projects), which are managed and financed by the individual state governments (Section 4.3.1).

Table 4.1. Seed certified (tonnes) in Nigeria, 2005–2009. Source: National Agricultural Seed Council.

		2005	2006	2007	2008	2009
Maize hybrid	ADPs	0	0	0	0	0
	Companies	1,386	2,948	1,137	2,641	3,150
	NGOs	0	0	0	0	0
	Total	**1,386**	**2,948**	**1,137**	**2,641**	**3,150**
Maize OPV	ADPs	680	1,006	1,244	3,437	264
	Companies	372	1,319	942	1,130	1,429
	NGOs	0	0	0	0	89
	Total	**1,052**	**2,325**	**2,186**	**4,567**	**1,782**
Rice	ADPs	1,005	1,647	2,909	6,501	227
	Companies	415	1,108	2,591	1,806	936
	NGOs	0	0	0	0	21
	Total	**1,421**	**2,756**	**5,501**	**8,314**	**1,184**
Sorghum	ADPs	55	77	82	180	17
	Companies	241	642	117	2,186	492
	Total	**296**	**718**	**199**	**2,366**	**509**

(*Continued*)

Table 4.1. Continued.

		2005	2006	2007	2008	2009
Pearl millet	ADPs	77	130	161	707	175
	Companies	0	0	0	295	764
	Total	**77**	**130**	**161**	**1,002**	**939**
Wheat	ADPs	22	40	70	250	0
	NGOs	0	0	0	0	93
	Total	**22**	**40**	**70**	**250**	**93**
Cowpea	ADPs	70	105	92	995	49
	Companies	24	37	62	84	49
	NGOs	0	0	0	0	17
	Total	**94**	**142**	**154**	**1,078**	**114**
Soybean	ADPs	41	257	277	417	15
	Companies	61	78	138	129	191
	NGOs	0	0	0	0	34
	Total	**103**	**335**	**415**	**546**	**240**
Groundnut	ADPs	47	122	121	392	26
	Companies	15	14	23	12	47
	Total	62	136	144	404	73
Cotton	ADPs	4	47	94	301	15
	Total	**4**	**47**	**94**	**301**	**15**
Sesame	ADPs	8	13	19	233	3
	Companies	0	0	0	12	0
	Total	**8**	**13**	**19**	**245**	**3**
Total		4,525	9,589	10,079	21,715	8,101

There is a small organization of Nigerian private seed producers, the Seed Association of Nigeria (SEEDAN), with headquarters in Zaria. It has slight impact on policy or the industry. All seed companies are subscribed members.

Seed cluster. As seed expert Robert Tripp noted, several seed companies often locate near each other, to facilitate information flow (Tripp, 2001). Several seed companies are based in the small northern city of Zaria, in Kaduna state: including Premier Seed, Alheri Seed, Nagari Seed, Maslaha Seed and the ADP seed activities for Kaduna state.

The seed companies clustered around Zaria benefit from the existing packaging industry.

The seed companies are conveniently located near the National Agricultural Extension and Research Liaison Service (NAERLS) of Ahmadu Bello University (once the largest in Africa south of Cairo). The university is also home to the Institute for Agricultural Research, where companies may get breeder seed and foundation seed. The NAERLS is a competent media programme at the university, producing radio and television programmes as well as written material in five languages (English, Hausa, Yoruba, Igbo and Nigerian Creole). The north-west regional headquarters of the Seed Council are also in Zaria, where small seed companies can process their seed.

4.1.3 Policy

A change in policy on import tariffs for cereals may have affected seed enterprises. In 2004 the government imposed a 130% levy on rice imports to protect Nigerian rice farmers, which stimulated farmers to buy seed. The high import tariff was lifted during the 2008 food crisis for 6 months. After the crisis, the tariff was replaced with a much lower one (32.5%) (Ibrahim Bamba, personal communication). Table 4.1 shows a surge in seed produced in 2008, in response to rising grain prices, but it fell again in 2009, when the tariff on grain was much lower. High grain prices stimulate demand for seed (see Section 10.1.5). Certified seed production for all crops is still low in Nigeria, although it increased sharply from 4525 tonnes in 2005 to 21,715 tonnes in 2008, before plummeting to 8101 tonnes in 2009 (Table 4.1).

The changes in certified rice seed production were similar. The authors believe that the policy on import tariffs helped drive up seed production due to higher demand by farmers, as seed demand is driven by paddy profitability, supported by high tariffs. Unfortunately, increased production was not accompanied by improved processing facilities, which could discourage further increases in production and the use of certified seed.

There are no taxes on seed or on basic food production, although supermarkets and exports are taxed. The government supports the formal seed sector through the National Agricultural Seed Council (NASC), which provides seed producers with foundation seed and quality control (Section 4.3.2). The NASC is federal. The federal and state ministries of agriculture often buy seed from companies to give or sell at subsidized prices to farmers. This not only discourages companies from advertising and marketing, but late government payments, outstanding after 2 or 3 years, also tie up their operating capital.

It is government policy to encourage the production and use of certified seed of improved varieties. Each state funds agricultural extension through its agricultural development project (ADP), which also promotes formal seed.

The National Agricultural Seed Decree No. 72 of 1992 by the (then) military government established the NASC and created its board of directors. The law is vague regarding plant breeders' rights. The Crop Variety Registration and Release Committee is part of the Seed Council. The Committee may take 2 to 4 years of trials (on-station and at several locations on-farm) before releasing a variety in Nigeria. Under the Decree (now the Seed Act), only released crop varieties can be marketed. While certification is voluntary, all seed to be sold must be 'truthfully labelled' (i.e. name of the variety, moisture and germination percentage). Certification is compulsory for all breeder seed, foundation seed and hybrids.

Seed certification. The 1992 decree established the Seed Certification Department of the NASC as the seed certification agency. The NASC assigns a seed certification officer to each seed company (or sometimes one officer for a group of small companies) while one to three seed officers are assigned to each of the ADPs (state governments) that produce seed. There are few complaints from seed producers about the arrangement, which streamlines certification.

In spite of this, and for reasons not entirely clear, there are still some cases of seed contamination. Even some of the breeder seed, which should be pure, is contaminated with other varieties. Mixture gets worse as seed moves down the chain to foundation seed and certified seed.

Regional harmonization. In 2009, new agreements with ECOWAS (Economic Community of West African States) led to two changes. First, it will speed up the release of new varieties, as any variety released in one member country may be sold in others. Secondly, all seed must now be certified. Nigeria realized that it would be impossible to certify all seed and argued against it during negotiations, but it was outvoted. Companies can no longer sell uncertified but truthfully labelled seed, so the formal sector must decide what to do with legislation that is impossible to meet. The same is true for Uganda (see Chapter 10, this volume).

4.1.4 Seed use and demand

Certified seed use. Between 5 and 10% of the national seed requirement comes from certified seed, the rest from the informal sector or farmer-saved seed. Except for a few crops, most varieties being planted now are improved varieties, although many have been planted by farmers for a long time without renewing the seed.

In-between categories of seed. Until recently seed could be sold without being certified in Nigeria, as long as it was truthfully labelled. The ECOWAS agreement will make uncertified, truthfully labelled seed illegal for 11 major crops. The Seed Council no longer tolerates companies selling uncertified seed. Farmers, however, may continue selling seed to other farmers (Box 4.2), in small amounts.

Problems. There is little irrigation so almost all seed must be grown in the rainy season and stored for the next year. Bank loans are hard to get and expensive.

This is in sharp contrast to Mali, where the government has created favourable financial conditions to support the agricultural sector, including the seed sector (Section 5.1.3).

Demand and varieties. Most maize varieties come in a yellow version for animal feed and a white one for people to eat. There is both upland and lowland rice. Northern Nigerians prefer sticky rice, to pound into a thick ball (a bit like mashed potatoes), while southerners like grainy rice. The north produces white cowpeas to eat locally, and brown ones are preferred by the southern market. Northerners like sweet cassava, while southerners eat dishes made from cassava flour, which can use sweet or bitter cassava.

Seed exports. One Nigerian company (Premier Seed) exports maize seed (hybrids and open-pollinated varieties – OPVs) within West Africa. The Nigerian government encourages exports by placing no tariffs, duties or other restrictions on companies exporting seeds.

Reasons to buy seed. Most of the formal seed is hybrid maize, even though maize is not Nigeria's biggest crop. Demonstrations by seed companies, NGOs and ADPs have convinced farmers of the advantages of using certified seed. In the case of maize and sorghum, which are raw materials used in the brewing and confectionery industries, farmers get a premium price for certain varieties, which encourages the use of certified seed. The same is happening with some rice varieties which are supplied by farmers to rice processing companies. Farmers buy seeds because of perceived benefits of a secured market, higher yield and resistance to pests, especially the parasitic striga weed, and when they cannot maintain the viability of their own seed.

Case studies. The following case studies describe five of the 13 registered seed companies in Nigeria, one NGO (Share Foundation) and a government structure (the ADPs). We start with the NGO because it focuses on foundation seed production, the source of the seed for other enterprises producing certified seed.

4.2 The Share Foundation

4.2.1 History

In December 2005, the Share Foundation harvested some 240 tonnes of valuable foundation rice seed. On 2 January 2006, that seed was all destroyed in a fire. Share never recovered from the loss, but is still trying.

The head and founder of the Share Foundation is Dr Sam Fagade, a professional agronomist and former head of Nigeria's rice research programme, who has also worked at IITA (International Institute of Tropical Agriculture) and AfricaRice. The case of Faso Kaba in Mali (Section 5.2) describes another person with contacts with research who was able to become a seed entrepreneur.

In 2001, Fagade started an NGO dedicated to teaching farmers about rice growing in Oyo State, south-west Nigeria. But in 2005 Share Foundation got an opportunity to lease 100 hectares of irrigated land and to obtain a loan for 50 million Naira ($333,000) to grow rice seed.

Fagade and his son Lekan, an agricultural engineer, bought two Indian tractors, took over a cluster of buildings on a state farm, and planted 100 hectares of rice seed.

Harvesting so much rice was a challenge. Irrigated rice is harvested at the end of a long dry season. The tonnes of rice were tinder dry. The Fagades took a break for Christmas. On the second of January Fagade got a phone call. The rice seed was on fire.

To his credit, Fagade never gave up. For one thing, he refrained from withdrawing the remaining 30 million Naira ($200,000) of his loan. He still had the rest to pay off, but he would avoid getting deeper in debt.

Dr Sam Fagade and a sturdy, Chinese harvester-thresher.

4.2.2 Structure

Diesel to run the irrigation pumps was a major expense. So, after the fire, Fagade turned to growing foundation seed in the rainy season, on smaller amounts of land, 5 to 6 hectares, about 20 tonnes of seed. The Share Foundation is one of nine outgrowers, producing foundation seed for the National Agricultural Seed Council. Sam Fagade's son, Lekan Fagade, is the farm manager, supervising two workers. They still have both tractors, a wagon, a boom sprayer, a newer, sturdier harvester-thresher, besides a small campus of shoddy, crumbling buildings, left over from the days when the land was a state farm.

Although the Share Foundation has close ties with IITA, AfricaRice and national research programmes, its closest link is with the Seed Council, which inspects the rice crop four times a year, provides the breeder seed and then buys the harvested foundation seed in bulk, hauls it to Ibadan (an hour's drive away) to have it cleaned, conditioned, packaged and distributed to producers of certified seed.

4.2.3 Cash flow

The Share Foundation has managed to get small grants of $3000 to $5000 each, from the FAO, and grow enough foundation seed to cover costs, but Share has still not repaid its loan and Fagade says he has not paid his staff in a year.

In 2009, it was awarded $150,000 from AGRA (Alliance for a Green Revolution in Africa) to produce rice seed, hybrid maize and soybeans, and it is now looking forward to a better year.

4.2.4 Marketing

For now its only marketing is through the Seed Council, which buys all of the Share Foundation's seed.

4.3 The Agricultural Development Projects (ADPs)

4.3.1 History

The ADPs (agricultural development projects) started in six states in 1975 as a World Bank funded activity. Between 1981 and 1984, the ADPs expanded from six pilot projects to statewide projects in three states (Kano, Sokoto and Bauchi). By 1992 there was an ADP in each state of Nigeria (Cheema *et al.*, 1994). These were integrated rural development projects which did everything from road building to agricultural extension. As old states divided, the new, smaller states also added ADPs. Now each of Nigeria's 36 states has an ADP, as does the Federal Capital Territory.

Every state in Nigeria has its own agricultural development project, which often inherited infrastructure from donor and government projects. Most have continued producing subsidized seed, albeit not for all crops.

Just as the ADPs were starting, from 1975 to 1980, the FAO supported the National Seed Service (now the National Agricultural Seed Council) with a seed project funded by the UNDP. The second phase (1981 to 1985) was funded by the Nigerian federal government. This seed project created seed processing and testing facilities in the cities of Ibadan (south-west Nigeria) and Zaria, in the north. Many seed officers were trained and seed certification was upgraded with equipment and training of laboratory staff (Usman, 1994). At first the ADPs were asked to produce seed; after all, there were no private seed companies in Nigeria until AgSeed started in 1984 (see Section 4.4). By 1992 the ADPs were producing massive amounts of seed, enough to plant 8–15% of Nigeria's crops, and private companies were complaining of unfair competition; ADP seed was so heavily subsidized that sometimes it was cheaper than grain in the market (Joshua, 1994). In fact, ADPs produce no hybrid maize, which by government policy is the responsibility of the private sector. ADPs produce seed that is less profitable, including rice (Table 4.1).

In 2010 the 37 ADPs are still agencies of the state governments, only loosely connected to each other (i.e. there is no national head of the ADPs) (Madukwe *et al.*, 2002). Many ADPs still produce seed, but not all. We visited the ADPs in Oyo, Ondo, Ekiti, Kaduna and Kano states, which all produce seed, especially maize OPVs, rice, soybean, cowpea and groundnut.

The ADP buys foundation seed from the National Agricultural Seed Council, distributes it to outgrowers, buys back the harvest, conditions and packages the seed and sells it to smallholder farmers for 120 to 150 Naira per kg ($0.80 to $1.00) for open-pollinated maize, while private companies sell their hybrid maize seed for about 200 Naira ($1.33) per kg.

Not all ADPs were created equal. For example, the Kaduna ADP inherited massive administration buildings from the days when Kaduna was the British headquarters for all of northern Nigeria. The ADP there was one of the first pilot projects, endowed with road building equipment and a large seed processing plant. As a second example, Ondo State was divided into Ondo and Ekiti in 1991; Ondo got the old agricultural office buildings, but Ekiti got the biggest seed plant in Nigeria (perhaps in all of West Africa), which could process 3 tonnes of seed an hour, while other factories could handle no more than 2 tonnes (Shobowale, 1994).

Since the early days the ADPs have lost staff and funding. But they all survived, even after structural adjustment crippled the formal sector. For example, in Kaduna there were 3000 extension agents serving 400,000 farm families, but after 1997 there were just 300 extension agents for a growing population that soon reached 500,000 farm households. The Kano ADP nearly stopped producing seed for several years and only started again in 2004.

The ADP in Kaduna survived budget cuts by using its heavy road building machinery to win construction projects. Other ADPs got donor-funded development projects, allowing them to continue. For example, the Multinational Nerica Dissemination Project has been funded by the federal government and the African Development Bank since 2004 in six states. This allowed the six ADPs to increase the amount of Nerica rice seed they distributed. And all the ADPs had modest budget support from their states.

ADPs that had seed plants (Kano, Kaduna and Ekiti) were able to keep most of the equipment running, no small feat since by 2010 most of the machinery was 30 years old and too big, and spare parts were hard to find and expensive.

The ADPs were able to ride out political changes. For example, in 2003 the federal government started the Presidential Initiative for Increased Rice Production and Export (as well as initiatives on maize, oil crops and others), whose strategy included the sale of subsidized, certified seed to farmers. Seed was produced through private companies. The federal government bought the seed with a 50% down payment and distributed it to the ADPs, which were supposed to recover the other half of the price from the farmers and pay it to the seed companies. Most ADPs could not recover the money as the farmers complained that they did not have the money, or the seed arrived too late to plant or was not what they wanted. The ADPs simply gave it away, saddling the seed companies with bad debt, which was eventually settled by the federal government through the Seed Council, who had to pay the seed companies back the money that could not be recovered. Besides this federal programme, the ADPs kept producing and selling their own subsidized seed to farmers.

4.3.2 Structure

Management. The ADPs are well managed. When we visited them in Oyo, Ekiti, Kaduna and Kano states in February 2010, the staff would quickly arrange for us to meet their programme manager, a person who had held the post for 10 years. The manager would speak knowledgeably about seed issues in his ADP. Table 4.2 shows the amount of certified seed raised by the ADP outgrowers in Kaduna, a major maize-producing state. Kaduna farmers buy more hybrid maize seed from Premier Seed Company than any other state. Note that most of the seed in the table is rice, suggesting that ADPs are an important source of rice seed, which may not be as profitable for private companies.

An ADP seed unit is typically managed by an assistant director of seeds, who answers to the director of technical services, who is under the programme manager (the head of the ADP).

Within the seed unit, the assistant director of seeds supervises an assistant seed officer, typically an agronomist, who oversees internal quality control. The state ADPs are divided into three or four zones, which have one manager, technical officer and seed officer per zone. In each local government (municipality) ADP extensionists facilitate the seed work (e.g. finding outgrowers and training farmers) (Box 4.1).

Table 4.2. Certified seed produced (tonnes), Kaduna ADP outgrowers, 2008.

Rice	236
OPV maize	102
Soybean	25
Sorghum	21
Cowpea	15
Total	399

Structure. The ADPs buy back the seed from outgrowers, although sometimes they may not buy it all. At least one ADP gives per diem to the inspectors from the Seed Council. This is not legal, but it does suggest that the ADPs want to be inspected, on a timely basis.

During the Presidential Rice Initiative (2003–2007), the Kaduna ADP had about 140 outgrowers. After the initiative ended, the ADP reduced it to 37, but they were still growing tonnes of rice, OPV maize, sorghum, cowpeas and soybeans. Outgrowers often sell uncertified seed to their neighbours (see Boxes 4.1 and 4.2).

The ADP hauls the seed to the seed factory, but if it does not have one it may use one of the Seed Council's factories. The ADP cleans the seed, bags it in 100 kg sacks and pays the outgrowers 150 Naira ($1.00) per kg, usually within 3 months of harvest.

The ADP stores the seed and then near planting season conditions it, dusting it with fungicide and insecticide, and bags it in smaller plastic bags of 5 kg to sell to farmers at just 150 Naira per kg. The ADP pays for all of the processing and transaction costs.

Links and partnerships. The National Agricultural Seed Council makes the whole system work. It holds an annual planning meeting in December–January, one in the north (Zaria) and one in the south (Ibadan), with the ADPs, seed companies, research institutes and some of the outgrowers. The ADPs and the companies submit 'indents' (estimates of foundation seed

Box 4.1 A Nerica seed grower in Kaduna.

Mallam Nasiru, in Kinkiba village, Kaduna state, started growing seed in 1992, to help other farmers. In 2000 he started growing Nerica 1 and 2 (rice). He also grows sorghum and maize seed. He started with less than 2 kg but in 2009 he sold 30 to 40 bags of rice (of about 80 kg each), 30 of maize and eight of Nerica 2 to the Kaduna State ADP.

The ADP pays the cost of inspections. Mallam Nasiru wasn't even aware there was a cost. He assumed that the certification officer was part of the ADP.

He is also in the PVS (participatory varietal selection) group sponsored by the ADP, where he tries new varieties and encourages his neighbours to see them. For example, he is now trying Nerica 8. Sometimes he also sells seed to his neighbours, for 150 Naira ($1.00) a kg, the same price as the ADP. The seed the neighbours buy is not certified, but is excellent seed.

Box 4.2 Seed growers of Bida Bidi.

Matina John is a woman farmer in the village of Bida Bidi, on the highway east of Zaria. We met her by accident, looking for women who had seen rice seed videos. She had not seen the videos, but she did produce seed, even though her extension agent was unaware of it.

Mrs John started in 2000, producing about 15 bags of 100 kg of maize and 25 bags of paddy rice. Encouraged by the high yields of improved varieties she began producing seed for sale. She buys the seed from ADP outgrowers who live nearby. Her outgrower neighbours are people like Mallam Nasiru (Box 4.1), who are selling seed that is good enough to be certified, although strictly speaking it is not certified. She pays them in cash and changes the seed every 2 years. She removes the off types.

She does get visits from extension staff, who advise her on growing crops. She sells the seed on her own. She stores the seed in new bags and sells it in those bags. In 2009 she sold five bags of rice, nine of soybean, 15 of maize, six of cowpea and ten of sorghum, at Naira 7000 each ($47). That is, she is selling at about half the price of certified seed, and some farmers are buying fairly large amounts of seed from her. She sells the seed to friends and neighbours, advertising only by word of mouth and her good reputation.

Her neighbour, Bello Yarima, is 52 years old and started farming in 1975. In 1990 he started selling seed. Unlike Mrs John, Mr Yarima uses certified seed as his source seed; he buys it from agents of seed companies, who have shops in the city of Kano. He buys new seed every 2 years. He has done well enough to buy 4 hectares of land in 1995, and to rent 2 more hectares in 2009. He has a special room in his house for storing seed. He only sells to farmers, never to the ADP or institutions.

He taught himself to grow vegetable seed. He asked vegetable seed growers in the market how to grow the seed, and began reproducing it himself. He measures it in empty tins of condensed milk. In 2009 he sold ten bags of maize, six of soybean, four of sorghum, five of groundnut and 30 to 40 milk tins of tomato, 30 to 40 tins of pepper and ten to 20 tins of onion.

He generally does not package his seed. Buyers bring their own bags, or he will sell them one. Vegetable seed buyers bring their own tins.

Both of these villagers are producing seed of improved varieties which are relatively recent, but which are already known in the area. In other words, the neighbouring farmers who buy this seed do so to get a supply of good seed at an affordable price, not to acquire a variety that is new to the area.

needed) at these meetings. Based on these and seed used from the year before, the Seed Council determines how much foundation seed to distribute.

The National Agricultural Seed Council gets breeder seed from national and international research institutions like IITA and AfricaRice, and distributes it to its own network of outgrowers in each of its five zones, inspects the fields, collects the harvested foundation seed, cleans it, stores it and conditions it before distributing it to the ADPs, private companies, NGOs and other small seed enterprises. The Seed Council is thus the main channel for getting new varieties from research to Nigerian seed enterprises, which is key to maintaining demand for seed.

Quality control. The Seed Council has one to three seed certification officers seconded to each ADP. They have their office in the ADP and visit fields with the ADP's seed officers. This joint quality control team inspects outgrowers' fields and the ADP pays the Council some amount per hectare per year for inspections. Fields are rejected if they are too weedy, diseased or not adequately rogued or if they have mixtures of other varieties. The seed certification officer from the Seed Council works with the ADP seed officers, inspecting each outgrower four or five times per year. At the last visit they collect samples of harvested seed for laboratory tests.

4.3.3 Cash flow

The ADPs pay cash for the foundation seed using funds from their state governments. The state budgets are supposed to be approved in December (for the following year) although sometimes there are lags and funds may still not be released in the early months of the year. But the ADPs are usually able to pay for foundation seed when they need it, sometime between March and June.

The ADPs advance their outgrowers fertilizer, foundation seed and herbicide, deducting these costs without interest from the amount paid at harvest.

Whereas the public sector produces most of the certified rice seed in Nigeria, private companies claim they could also cover this.

Private companies complain of unfair competition with the ADPs. There is a bit of truth to that; perhaps companies would be able to produce more OPV maize and rice, for example, if the ADPs were not producing it. But the ADPs produce no hybrid maize seed, leaving that profitable product to the companies. And there is also collaboration between companies and ADPs that do promote modern varieties, driving up demand for the companies' products. The companies and the ADPs attend about three annual training courses from the National Agricultural Seed Council on seed production, quality control and postharvest management. We cannot put an exact figure on the profitability of the ADP seed enterprises, but they do receive several subsidies, e.g. equipment inherited from donor-funded projects, and an operating budget that comes from their state government. Because the ADPs have to turn over gross receipts from seed sales to their state treasury they pay little attention to profit as a motive.

When seed companies receive big orders to deliver seed to the federal government or a donor-funded project, the companies may poach seed from outgrowers of the ADP, which is generally tolerant of it, telling farmers, 'We're just glad you sold your seed.'

4.3.4 Marketing

At the end of each year the ADP assesses its seed stocks, and leftover seed is sold as grain. Seed is fumigated in storage, to kill insect pests (especially weevils), but the seed is only dusted with fungicide and insecticide just before sale.

Almost all of the ADP's seed is sold directly to farmers. The ADP announces seed sales on the radio. The prices are posted on bulletin boards in ADP office compounds and the seed is also sold there. Hundreds of farmers enter at the appointed day and pay cash for the low-cost seed.

ADPs keep trust by maintaining reasonable quality. The certified seed is grown by experienced, trusted outgrowers. Their biggest competition is not from seed companies, but from farm-saved seed and from informal seed bought in grain markets (Box 4.2).

As an enterprise which is decades old, and has never changed its name, the ADP is well known and needs little advertising, other than the radio announcement of sales. Any other advertising is through the extension agents.

4.4 Premier Seed Nigeria

4.4.1 History

Premier Seed Ltd in Zaria is Nigeria's largest seed company. It was founded in 1989 by Chief Olusegun Obasanjo, former president of Nigeria, as Parental Line Seed Limited, which merged with the Pioneer Hi-bred Seed Company of the USA in 1990. In 1992 Pioneer bought an older company, Agricultural Seed Nigeria Limited (AgSeed), founded in 1984 by the Leventis Foundation. In 1994 Pioneer pulled out of Nigeria, no doubt frustrated by low demand for seed by smallholder farmers and the collapse of large-scale farms, as well as the competition from ADP-subsidized seed, and Pioneer's Nigerian holdings became Premier Seeds, owned by Obasanjo, who has various other agricultural companies.

From the beginning Premier specialized in hybrid maize seed; one of the first directors was an IITA maize breeder. Premier produces its own inbred lines for hybrid maize. Premier also produces seed of OPV maize, rice, soybeans, cowpeas, groundnuts, sorghum, pearl millet, cotton, okra and various local leafy vegetables, such as *tete* (*Amaranthus hybridus*), *soko* (*Celosia argentea*) and *ewedu* (*Corchorus olitorius*). Premier imports seeds of tomato, watermelon, cucumber, carrot, onion, sweet pepper, cabbage and lettuce.

In 2003 Premier obtained release of two new hybrid maize varieties, Oba Super 1 and Oba Super 2, and in 2009 it obtained release of six more (three yellow and three white), all products of its own research. Premier has registered these hybrids with the Crop Variety Registration and Release Committee, and has sole possession of the parent lines. The law protects the right of Premier against others producing their hybrids. In 2008 the varietal release committee set up a 'Stakeholders' Committee on Access and Benefit Sharing' to discuss the IPR (intellectual property rights) period to be granted to institutions or private breeders for any developed crop varieties. The period is yet to be determined but will soon be. Premier is the only seed company in Nigeria to have developed and obtained release of any crop variety.

4.4.2 Structure

Management. Premier Seed sees itself as a modern, Nigerian corporation, doing solid research and breeding hybrid maize. Premier sells seeds of other crops, but hybrid maize is 75% of its field crop seed (Table 4.3).

Some years Premier buys some foundation seed from the Seed Council, but Premier uses 3000 outgrowers to produce most of its foundation seed (including inbred lines of maize) and certified seed. After cleaning the seed, Premier stores it in 100 kg bags in a large, clean warehouse at company headquarters. The scent of phostoxin (insecticide) in the air suggests zero tolerance for weevils. Inbred lines and foundation seed are held in a separate warehouse to avoid mixing them with certified seed. Before sale, Premier conditions the seed with fungicides and insecticides, and packages it in 2 and 5 kg bags for the smallholder market.

Table 4.3. Seed produced (tonnes), Premier Seed.

	2005	2006	2007	2008
Hybrid maize	928	1,048	943	1,191
OPV maize	101	–	25	78
Rice	259	187	178	172
Sorghum	4	7	5	29
Soybean	14	34	12	63
Cowpea	3	5	1	7
Total	1,341	1,818	1,159	1,540

Premier employs 66 people. The workforce is so stable that six were part of the old AgSeed Company. Premier participates in training with the Seed Council and also encourages its employees to seek university degrees. The managing director, the head of research and two other staff have PhD degrees and other key people are university graduates.

Infrastructure. Premier is divided into administration, accounts, marketing, production, research, processing plant and laboratory. In 2009 Premier began building a new plant at its headquarters in Zaria. The buildings are now partially complete and the equipment is yet to be installed.

Equipment. Premier is still using the original equipment of AgSeed, including a seed cleaner (three storeys tall), an equally large seed conditioner and industrial-sized warehouses, all in excellent condition after nearly 30 years of use.

Everything in order. Premier's tidy seed germination test.

Links. Premier proudly claims links with IITA, AfricaRice, CIMMYT (International Maize and Wheat Improvement Center), the 'universities of Nigeria' and other research institutions, although its ties with IITA are indispensable as a source of new maize lines.

The managing director of Premier is also the President of the Seed Association of Nigeria (SEEDAN), and built a small office building for them at Premier's headquarters. The managing director also represents the Seed Association on the Seed Council's board of directors.

Premier resents the ADPs' cheap prices ('the ADPs are our number one enemy') but the ADPs also promote hybrid maize throughout Nigeria, relieving Premier of having to invest in promoting its main product. Premier enjoys subsidized training and quality control from the Seed Council, and pays little or no tax on imports. But Premier complains that the government and ADPs should do much more. Premier is working on two projects with IITA and the Bill and Melinda Gates Foundation: drought tolerant maize for Africa, and the Tropical Legume Project.

Quality control. Premier's laboratory is clean and with just the right amount of equipment (some scales, a germination chamber). Premier's laboratory has long trays of several hundred bowls, filled with soil, where the staff work, testing the germination rate of its seed.

There are Seed Council officers seconded to Premier. But Premier Seed thinks the national seed laws need to be revised, especially to strengthen varietal protection. No one else has copies of Premier's parental maize lines, which is a form of protection against the vague plant breeder's rights in Nigeria.

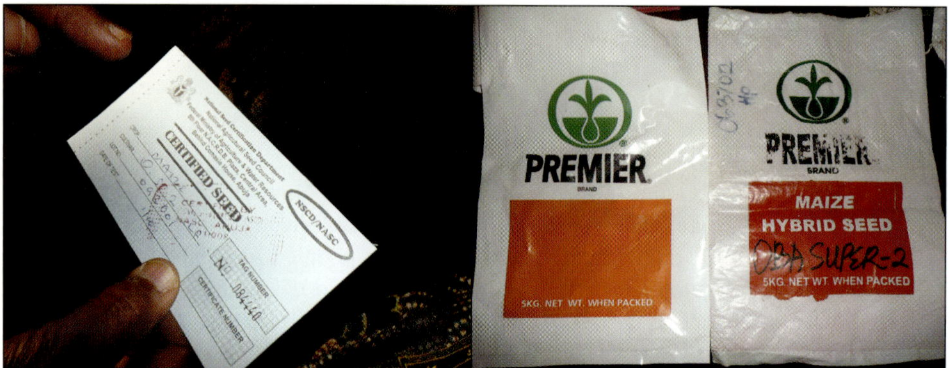

Ensuring quality is not enough. Counterfeited seed is put in fake bags (e.g. the one on the right with red label), as such undermining a company's reputation.

Premier also wants to see the law strengthened to make it easier to catch seed counterfeiters and to make their penalties stiffer. Premier loathes the seed counterfeiters. These crooks make fake bags that look almost exactly like the Premier bags. Counterfeiters also make bags of other companies, including imaginary ones. The counterfeiters fill the bags with any maize grain at hand, and coat it with a powder and sew the bag shut. They even add a fake certification label. With grain selling atw 60 Naira ($0.40) and commercial certified seed at 200 Naira a kg ($1.33), counterfeit seed is lucrative. In 2006 Premier prosecuted one group of counterfeiters but there are others still at large.

4.4.3 Cash flow

Government resources are important, but a double-edged sword. The federal government still owes Premier for seed purchased for the Food Security Programme.

In 2009 the government promised to pay 40% of the costs of the new seed plant being built at Premier, but has still not disbursed the funds.

Like other seed companies, Premier Seed pays for everything in cash, and sells on credit, months later. While we were visiting them, the managing director got a call promising to repay a 17 million Naira credit ($113,000). There was an obvious feeling of relief around the table.

To finish building its seed factory, Premier spent the money it was keeping in savings to pay for its inputs. After months of negotiations for a bank loan, with no success, Premier had to seek funds from the company's owner, the ex-president.

At planting time, Premier loans fertilizer and seed, at cost, on credit, to its outgrowers. It then deducts these costs from the payments made after harvest. The outgrowers almost always honour their contracts with the private companies. This is probably because many of the ADPs encourage so many outgrowers that there is a surplus of outgrowers. The ADPs cannot always buy back all of the outgrowers' harvest. In Nigeria formal seed is a buyer's market and the farmers are the ones who worry that the company or the ADP may not buy back all of their harvest, and not the other way round.

Premier does not mind competition from respectable seed companies, and is proudly aware that over the years several multinational seed companies have come to Nigeria and failed. For example, Pioneer Seed left Nigeria for several reasons – because they got tired of the competition from the ADPs, due to the challenging economic situation in Nigeria at the time.

4.4.4 Marketing

Government agencies used to take most of the certified seed, but that has recently changed. Premier currently targets smallholder farmers and international NGOs, both of whom tend to pay for their seed on time. Premier sells large volumes to the government, especially up to the end of 2006, but payments are up to several years late and have led to devastating cash shortages; 2006 was a peak year for sales, after which government programmes began buying much less seed.

In the future Premier wants to concentrate on a dedicated network of seed dealers and is grooming three dealers in each of the 774 local government areas. By 2010 Premier hopes to have over 2000 seed dealers. Premier also has its own shops in some major cities, such as Kano, which sell seed directly to farmers and to dealers. Also in Uganda medium-sized companies increasingly focus on the dealer and retailer market (Section 10.2).

Like other Nigerian seed companies, Premier does almost no advertising through the media. This is partly out of a sense of false economy, partly out of habit from the days when government could be counted on to buy seed. Premier thinks that the ADPs should do more to promote hybrid seed. Most of the seed companies in Nigeria show an amazing lack of interest in advertising. Compare this with the aggressive advertising by an enthusiastic farmers' group in The Gambia (Section 7.3). Premier does set up some 50 demonstration plots per year, and holds a field day at each one, but only nearby farmers attend, a tiny fraction of their total market.

4.5 Nagari Seed Nigeria

4.5.1 History

Nagari Seed Nigeria Ltd, in Zaria, started in 2000, but its history really began in 1984 when the current managing director of Nagari, A. Boman, was the production manager of AgSeed, which became part of Premier Seed in 1992. In 1996 Mr Boman resigned and moved to Alheri Seed, which he managed until 2000, when he resigned and joined the newly formed Nagari Seed.

The small private company registered with the Corporate Affairs Commission in September 2000. Mr Boman had also previously worked with the Kaduna state government, which loaned him a small two-room office in an old ADP building in Zaria.

That same year, the company got its first four inbred lines of parent seed from IITA to produce hybrid seed. In 2001 the National Agricultural Seed Council gave it foundation seed for OPV maize. The company came out with its first hybrid maize seed in 2004.

In 2004 Nagari Seed rented a large farm, over 80 hectares, for producing seed. But it was not feasible. They were seed experts, not farmers. So the company shifted to outgrowers.

At first, it used the Kaduna state ADP seed processing plant at Magana (rented out to Alheri Seed), but later Nagari Seed switched to the NALCO plant (formerly used by the defunct UAC Seed Ltd) in Zaria. About 95% of its business is in maize (25% hybrid in 2005, increasing to 42% in 2009), but it also sells some rice. It offers some courses for its outgrowers, through WASA (West Africa Seed Alliance), a USAID-funded project.

4.5.2 Structure

Nagari Seed has a managing director, a secretary, a production manager, a marketing manager, a factory supervisor and an accounts person. It has 335 outgrowers, but does not use them all every year. In Nigeria the supply of outgrowers exceeds demand. In 2010 the company used 220 outgrowers, because in 2009 'the market was not encouraging, so this year we feel we need to cut down production'. The Seed Council has an officer assigned to Nagari Seed (whose salary and operating expenses are paid by the Council), who inspects seed in outgrowers' fields and in the factory. The company stores its seed in a rented storehouse.

It buys its seed bags from a manufacturer in Kaduna (who must be paid in advance). Two kg bags are the most popular for this smallholder market, but Nagari Seed also makes 5 kg bags.

Mr Boman feels that government policy is inconsistent. He would like fertilizer prices to be subsidized, so farmers would use more and get bigger yield increases from hybrid maize. Actually, fertilizer is subsidized in Nigeria, but because of distribution problems many farmers never get it. Mr Boman would also like the government to stimulate seed demand more. He knew about Premier's case with the seed counterfeiters and sympathizes with Premier. Fake seed hurts everyone producing certified seed.

4.5.3 Cash flow

The company advances foundation seed and fertilizer to its outgrowers. If it has the money, it pays the outgrowers 40% after cleaning their seed. 'We only pay for clean seed. We need to see how much it is first before we pay.' It pays the balance after processing. The price is agreed before planting. This seems risky, but it works most of the time in Nigeria; because there are so many outgrowers, the seed companies have the upper hand. Sometimes outgrowers need an advance to de-tassel the female parent line.

Nagari Seed has never sought a loan. Banks require collateral, which it doesn't have. At the end of every year it saves money in a bank account, to buy inputs for the outgrowers the following season and 'from our sales we pay the farmers'.

4.5.4 Marketing

The company markets through seed dealers, most of whom also sell agro-chemicals, but the bulk of sales is to individual farmers (Table 4.4). It also sells some seed through the ADPs. The seed dealers are unreliable customers, who take Nagari's seed on consignment and do not pay until they make a sale. Even then, some keep the money to run their business 'and we start fighting. One dealer still has N 400,000 ($2670) of my money.' At the end of the season, some seed dealers want to keep the seed through the dry season, but after that the seed is often weevily and the germination rate is low. Since the dealer never paid for the seed, they simply return it to Nagari Seed, which salvages such seed by selling it to cattle ranchers to plant forage maize. In spite of troubles with the dealers, the company sells 80 to 85% of its seed through them, to avoid the government, 'which ties down your money for the next 5 years'.

Table 4.4. Clients of Nagari Seed.

	2005	2009	2015 (predicted)
Individual farmers	2	1	1
NGOs	1	2	2
Farmers' cooperatives	3	3	2
Agro-dealers	4	4	3
Government	–	–	–

Ranking assessment by senior management of seed enterprise, 1 being the most important.

Nagari Seed has some slots on the radio, through the ADP and the NAERLS, a national extension service, headquartered in Zaria. The NAERLS has two radio stations and allows the company to air spots at a low cost. Help like this from federal government agencies allows seed companies to survive in the face of subsidized seed production by the ADPs, which are state government agencies, not federal.

Various seed companies are clustered around Zaria, where the government allows them to air spots on its state radio and TV at low cost.

4.6 Terratiga

4.6.1 History

Terratiga Ltd was formed in 2000, based in Kwanar Diwaki district, 16 km from Kano. Their main business is producing irrigated vegetables, herbs and spices for upmarket hotels and supermarkets in Abuja and Lagos. Vegetables were Terratiga's original business. Seed is a much smaller activity, which started several years later.

4.6.2 Structure

Terratiga rents 116 hectares from individual farmers for seed production, of which it uses 8 hectares for irrigated rice, 2 hectares irrigated land for groundnut seed and 100 hectares for upland (unirrigated) sorghum seed.

Terratiga gets its foundation seed from the Institute for Agricultural Research, Zaria, grows its own certified sorghum seed, supervised by the NASC man in Kano, and sells it to farmers, who produce high-quality sorghum for Nigerian Brewery, to make maltina (a sweet, non-alcoholic brew). The growers are connected to the USAID MARKETS project. Terratiga advances the farmers the seed and provides technical support for farmers to ensure grain quality.

One of the strange things about Terratiga is that they have no seed plant in Kano. They truck their seed to their company's seed factory in Lagos, 1000 km away, and then haul it back to Kano to distribute it to the farmers. This does allow Terratiga to avoid buying seed processing equipment, although it adds to transportation costs.

4.6.3 Cash flow

Terratiga is a subsidiary of AfriAgri Products, a trading company dealing in cocoa, coffee, sesame seed, ginger, hibiscus and other products. AfriAgri Products gets bank loans, and makes funds available to Terratiga to pay for inputs and to pay farmers.

The cost of the seed advanced to the farmers is deducted at harvest. The company buys fertilizer in bulk and makes it available to farmers in cash, at cost. This ensures that the farmers have access to good fertilizer, and avoids problems with repayment later. Farmers bring the harvest to Terratiga's factory near Kano and are paid a week later.

Terratiga produces mainly certified seed for farmers who supply high-quality sorghum for Nigerian Brewery.

4.6.4 Marketing

The sorghum farmers are a niche market created by the MARKETS project. Rice and groundnut seed is produced by contract for special customers. Otherwise, Terratiga does little seed marketing.

4.7 The Seed Project Company

4.7.1 History

The Seed Project Company Ltd never was a project, but was started as a private company in Kano in 2005 by two Nigerians, Lawan Gwadabe and Stella Thomas. Mr Gwadabe had always been in seeds, first as the director of the Research Department for the ADP in Kano and then as the chargé of commercial crops for the Ministry of Agriculture. When he left the Ministry he wanted to start a seed company. With just 70,000 Naira ($467) Gwadabe and Thomas began importing seed. They rented an office on the second floor of a commercial building. They still work in the same tidy, well-lit, two-room office.

The Seed Project soon began working with rice, cowpea and maize. They bought foundation seed from the Institute for Agricultural Research in Zaria and looked for outgrowers that Mr Gwadabe had known from his days at the ADP. Seed Project specializes in high yielding open-pollinated maize, with yields as much as 5 tonnes per hectare. Gwadabe and Thomas think OPV maize may open the way to selling hybrids in the future (see Section 10.2 for a similar line of thought by one of Uganda's leading seed companies, which, 15 years after its establishment, is producing over 2500 tonnes of seed). The Seed Project imports vegetable seeds by air, mainly from Italy and The Netherlands. The Seed Project claims that importing vegetable seed is still their biggest activity, although they are rapidly expanding into seed production (Table 4.5).

The Seed Project soon hired other staff, including 11 university graduates. The managers say that good short courses are unavailable, and that they do most of their staff training themselves. Their first few years were so successful they bought a 146-acre (58-hectare) farm to grow their own seed.

Table 4.5. Certified seed produced (tonnes), Seed Project.

	2005	2006	2007	2008	2009
Maize OPV	70	100	250	280	400
Rice	20	40	100	70	80
Soybean	–	–	–	5	60
Cowpea	0.2	0.5	5	2	15
Sesame	–	–	–	5	20
Carrot	–	–	–	–	8
Groundnut	1	0.5	2	5	7
Watermelon	–	–	–	–	3
Onion	–	–	–	–	3
Tomato	–	–	–	–	1.5
Cucumber	–	–	–	–	1
Sweet pepper	–	–	–	–	0.7
Okra	–	–	–	–	0.2
Total	91	141	357	367	599

4.7.2 Structure

Management. The managers of Seed Project say they have many ideas they could put in place if they only had more money, but they are happy to keep expanding on their own capital. 'Our goal was to reach a turnover of 150 million Naira this last year. We reached 143 million, so we were only off by seven.' Many small firms would be delighted to turn over nearly a million dollars.

There are 23 permanent staff members in three departments: marketing (seven), agronomy (four), which includes processing, production and research, and the rest in operations (office assistant, cashier, accountant, storekeeper, two drivers and four security guards). There are 50 day labourers, who work almost every day on the farm or in the factory.

The Seed Project imports watermelon seed from Senegal. They airfreight their other vegetable seed from Europe, often once a week, to make sure it is fresh. Seed Project has a person who clears the seed with customs in Lagos and puts it on the next flight to Kano. Most of the carrots in Kano are grown with seed imported by the Project.

Land. The 58-hectare seed farm includes 12 acres (4.8 hectares) of irrigated land, which the company plans to use soon to produce tomato seed. They are already producing onion seed, as well as maize and pearl millet, on the 58 hectares.

Outgrowers. The Seed Project uses 50 to 100 outgrowers to grow some 300 tonnes of maize (besides the 100 tonnes grown on the Seed Project farm) and 150 tonnes of rice seed. The Seed Project buys OPV maize seed from its outgrowers at 20% above the market price for grain. They are one of the few companies that pay flexible prices; most set a price at planting time. But, no matter how a company reaches a price, if it is high, farmers will be tempted to bring in grain from other fields to sell as seed. However, the certification officers estimate yield during inspection, so they know how much harvest to expect, to keep outgrowers honest.

Seed Project Ltd first imported and sold vegetable seed. It soon began growing rice, cowpea and maize seed and since 2009 expanded to vegetable seed.

Infrastructure. The Seed Project does not have a seed factory, but rents from the Kano ADP. The Project saves money by not treating most of its seed, claiming that the dusting of pesticides actually costs more than the seed.

4.7.3 Cash flow

About 30% of its business is with the government and 70% is with private farmers, because 'The government doesn't pay. You sell and you wait. That's why other companies are crumbling.'

Sometimes the Seed Project loans outgrowers money to weed their crops, but it tries to avoid loaning them too much. Mr Gwadabe has known some outgrowers for many years. In the first year the outgrowers basically loaned money to the Seed Project, planting a crop on their own, even paying for the seed, just on the promise of buy-back later. 'The farmers trust us.'

Mr Gwadabe says he manages finances by sheer discipline. 'You make a plan and you reduce your overheads.' Seed Project is prospering, but still owns very little, just one storeroom and the farm. They rent two other storerooms and everything else. They have never taken out a bank loan because 'you waste so much energy looking for it'. The interest is 24%, but with fees it ends up being 30%. Unless you earn 50% on your investment you end up working for the bank.'

4.7.4 Marketing

Seed Project has increasingly focused on reaching out to individual farmers and agro-dealers (Table 4.6). The mobile phone really helps. Seed Project puts its mobile phone numbers on all its packages: a trust-building, customer-friendly innovation. 'In the remotest villages they call with any problems,' and if the company cannot help over the phone, they send one of their own extension agents to visit them. It is one of the only seed companies that has extension agents.

Table 4.6. Clients of Seed Project.

	2005	2009	2015 (predicted)
Agro-dealers	–	1	2
Individual farmers	3	2	1
Government	1	3	5
Groups and Cooperatives	2	4	3
Projects and NGOs	–	–	4

Ranking assessment by senior management of seed enterprise, 1 being the most important.

The Seed Project gives credit to its agro-dealers, and some pay half in advance. 'You need to have a good strategy for collecting your money, and know how much credit to give them. They must pay off before taking more goods. After all, you prefer your goods to be with an agro-dealer than in your storeroom.' The strategy is working, because over 98% of the agro-dealers pay back their loans, a record for Nigeria. Seed Project has less than 4 million Naira ($26,600) in outstanding loans.

The Seed Project organizes field days for agro-dealers in three places in the state of Kano where they have planted their products. Project representatives also attend trade fairs to display their products.

The company also trains outgrowers and community organizations. Seed Project extension agents visit the local chapters (local, community-based groups) of RIFAN (Rice Farmers Association of Nigeria) one by one and give them training on planting density and fertilizer rates. They show farmers how to use a rope to calculate the land area and to figure out the amount of seed and fertilizer to use.

One year the Seed Project imported onion seed with a poor germination rate. Farmers complained and the company replaced the seed. It cost 1 million Naira ($6600) but it was worth it to keep their customers' trust. In another case, the Seed Project imported tomato seed from Top Harvest, an obscure Dutch company. Farmers and the ADP complained that the seed was of mixed varieties, so the Seed Project stopped importing that brand of seed.

4.8 Maslaha Seeds Nigeria

4.8.1 History

In Hausa, Maslaha means 'to ease, or to come to someone's aid'. Maslaha Seeds Ltd was started in Gusau, north-west Nigeria, in 2006 by a former senator, who had been a large cotton and grain grower since the 1970s. The senator had long been interested in trying new varieties and often visited research institutes looking for good seed. The senator and several other business people support Maslaha Seeds and sit on the board of directors. It registered with the Corporate Affairs Commission in 2007.

Maslaha Seeds hired a person from IITA as head of production and turned one of the senator's large farms into a seed farm. The first problem was finding a place to process the seed. The Seed Council suggested that it used their seed factory in Zaria, which required some minor repairs, which Maslaha fixed.

The first year, 2006, Maslaha Seeds produced over 600 tonnes of mostly rice and hybrid maize seed (Table 4.7). It was 'hell' hauling hundreds of tonnes of seed from the farm 100 km to Gusau, then making the 4-hour trip to Zaria to process the seed and then bring it back. But results from the first year were encouraging; the company sold 80% of the seed (mostly hybrid maize and relatively new rice varieties like Nerica 1), and decided to buy its own seed processing plant. It chose a Chinese brand, which was less expensive than European makes. Chinese engineers came to the site to help install it and Maslaha has been satisfied with the machinery.

Table 4.7. Seed produced (tonnes), Maslaha Seeds.

	2006	2007	2008	2009
Hybrid maize	196	446	540	784
OPV maize	135	418	255	959
Rice	255	469	412	487
Sorghum	2	6	149	369
Pearl millet	4	90	173	257
Soybean	35	53	87	85
Cowpea	3	51	1	45
Groundnut	2	3	0	5
Total	632	1,536	1,618	2,991

The second year, 2007, the company was glad to have reduced transportation costs, and more than doubled its output, to 1500 tonnes of seed. It began to produce significant amounts of pearl millet for the first time, in response to demand from farmers. That year it also started growing seed with outgrowers, having learned from the first year that it could not grow enough seed itself. Many of the outgrowers were farmers that the senator knew from his days in politics. The third year, Maslaha Seeds began conducting research, testing new varieties on its own farm.

4.8.2 Structure

Management. Maslaha Seeds now has 35 employees, in four departments: operations (production, processing and storage), research, finance and accounts, and marketing.

It has an optimistic vision of Nigeria and the future. It figures that Nigeria has 32 million hectares of cropland, requiring hundreds of thousands of tonnes of seed, 100 times the current output of the formal sector. It has a 5-year plan to grow, package and sell more seed. By 2012 it wants to produce 7000 tonnes of seed, about triple its current output. It also wants to start importing vegetable seeds.

Land. Maslaha Seeds uses one of the senator's farms, at first for seed production, but now just for research.

Infrastructure. In 2009, it began building four new large seed plants, with the federal government contributing 40% of the cost under public–private partnership. The seed factory in Katsina (including warehouses and offices) is now being roofed (February 2010), as is the one in Gwagwalada, near Abuja in central Nigeria: from there Maslaha Seeds hopes to serve southern Nigeria.

In 2008 Maslaha had less than 100 outgrowers, but Maslaha is expanding so rapidly that it now has nearly 1000, and in the future will only work with fairly large outgrowers, those who can produce seed from at least 5 hectares, in order to lower supervision and transaction costs.

At the beginning of the season Maslaha Seeds signs a contract with each outgrower, specifying the quantity of fertilizer to be used. It gives them the parent lines (for hybrid maize) or the foundation seed for other crops. After harvest the farmers bring the seed to a collection point, where Maslaha deducts their loan and pays them the balance.

Maslaha's affordable and functional Chinese seed factory.

Quality control. The National Agricultural Seed Council assigned an inspector to work with Maslaha Seeds, as with the companies described in previous sections. The Seed Council helps register and train outgrowers. New employees are trained on the job.

Linkages and learning. Maslaha has links with AfricaRice, IITA, CIMMYT, AGRA and other research organizations. In late 2009 the senator and the manager of the company went to India for a trade fair, and visited seed companies and researchers at ICRISAT, where they learned about high yielding hybrid pearl millet, which Maslaha Seeds wants to try in Nigeria.

4.8.3 Cash flow

Maslaha Seeds has obtained one loan, guaranteed by the UNDP at 8% interest, and is also interested in raising private capital. In most African countries, agricultural entrepreneurs struggle to obtain loans at an acceptable interest rate. Mali is one of the few countries where the government facilitated the creation of agricultural loans at 8–12% interest rate, the interest being lowest for farmers' associations and cooperatives (Sections 5.1.2 and 5.3.3). In Nigeria's neighbouring country, Cameroon, farmer seed producer groups reduce financial hardship by avoiding loans and by setting up revolving funds (Section 3.1.4).

4.8.4 Marketing

Maslaha has sold seed to the government, even though their slow payment can 'put you down on your knees'. Maslaha realizes that in the future they have to sell more to farmers (Table 4.8). They have agro-dealers, and also sell to farmers. Dealers ask for seed on credit. 'If you ask them to pay 50% in advance, they won't take the seed.'

Table 4.8. Clients of Maslaha Seeds.

	2006	2009	2015 (predicted)
Government	1	1	5
Individual farmers	2	2	1
Agro-dealers	–	3	3
Projects and NGOs	–	4	4
Groups and cooperatives	–	5	2

Ranking assessment by senior management of seed enterprise, 1 being the most important.

In their second year Maslaha sold 50,000 Naira ($333) worth of seed to farmers, and was encouraged that they came back the next year to buy 2 million Naira ($13,300) worth. 'With their own money!' Maslaha seems to have the right mix of connections (political and research), land, qualified management and capital.

4.9 Challenges and Strengths of the Seed Enterprises

Small-scale and low capacity utilization. At the national level, Nigerian farmers are planting modern varieties, but saving their own seed whenever possible. The use of formal certified seed (including hybrids) is relatively low in Nigeria. All of the seed companies are small (none sell more than 2000 tonnes per year).

Varietal mixtures. There have been several complaints that seed, especially lowland rice, is mixed with other varieties, even weed seed (wild rice). For example, in 2009, the USAID Emergency Rice Initiative commissioned 150 tonnes of lowland rice seed from three Nigerian companies. The farmers who received this seed later complained that the seed was mixed with other varieties and with wild rice. Lowland rice is especially susceptible to this problem, because it is grown in small pockets of wetland, year after year. So seed banks build up in the soil. To get a pure crop of only one variety, one would have to either grow just one variety for several years in a row, or spray herbicides at least twice before planting, until the seed in the soil was exhausted.

Companies do take a few steps to keep outgrowers honest. First, they retrieve the seeds as soon as possible after harvest from the outgrowers into the company's store. Second, they report the offending outgrowers to their community leader. And, third, if outgrowers continue to default they are blacklisted.

Foundation seed. There are some complaints from companies that foundation seed is of mixed varieties; sometimes even breeder seed is mixed. One company said it received such mixed foundation seed and had to rogue it nine times to keep the variety pure.

Equipment. Most of the enterprises (especially the ADPs) have ageing equipment.

Counterfeiters damage the whole seed industry, not just the companies targeted.

Plant breeders' rights need to be clarified.

Cash flow is a problem that all enterprises have to solve. The ADPs do not have a revolving fund for buying seed. Each year they must seek fresh funds from the state government. After selling seed at the planting season the receipts go back to the state treasury. Fortunately fresh money is usually available but there are occasional problems. For example, Ondo state was not able to grow rice seed in 2009 because funds were distributed too late.

Most companies find it almost impossible to get bank loans, and must save money to buy inputs and pay farmers. A few companies have owners with deep pockets.

Terratiga is able to manage cash flow by growing seed as a fraction of its activities, by growing it in the dry season and by belonging to a larger company with access to bank credit.

Alheri does get bank loans, although at a high cost. Nagari Seed has been able to avoid capital expense. They use an old, cramped office for free, and they rent their storage and processing facilities. Their only expenses are for labour, seed, fertilizer and bags.

The ADPs work because they are stable and local. The staff have been at their jobs for years. They are either from the state or have spent their adult lives there. They speak the local language(s), know the local people, identify with them, eat with them, pray with them and communicate. The state funds their seed activities every year both to boost the agricultural productivity of the state and to keep in the good graces of the farmers.

Through military and civil governments, peace and war, political dramas and the shifting support of international donors, for 35 years the ADPs have managed to continue selling seed at affordable prices to the farmers.

Businesslike. At Premier Seed everything from the offices to the factory floor is clean and in good working order. The machinery hums, the laboratory is appropriately used and everyone in the company is quietly at work. Terratiga, Maslaha and the Seed Project have a similar feel of competence and craftsmanship.

Marketing. This is the weak link. Nigerian seed experts are exasperated with companies that do not advertise, pay little or no attention to customer demand and expect the government to promote their seed. The seed companies do not always relate well with their agro-dealers, who complain of seed quality problems with certified seed (e.g. many immature or broken seeds and varietal mixture and mislabelling). The trend is for companies to try harder to make direct sales to farmers, with less interest in selling through agro-dealers.

The long seed chain. This starts when breeder seed leaves research institutes, goes to the Seed Council, which parcels it out to outgrowers, who produce foundation seed, send it back to the Council, which then distributes it to seed enterprises, who send it to more outgrowers, who send it back to the seed enterprises, who distribute it (often indirectly) to farmers. It is remarkable that such a long supply chain works at all. One is reminded of the old line about stunts on television (don't try this at home). We hesitate to recommend a chain with so many links to any other country;

yet, for the record, it is working in Nigeria. And it is a wonder that seed companies can even survive under competition with ADP-subsidized seed and a debt load created by government seed procurement programmes. But they do.

Among the various factors, successful seed enterprises all seem to: keep their overheads low; have a good product; be honest with outgrowers; and advertise their product. Nigeria's private seed companies are clustered around key research centres and are locked in a rivalry with the ADPs that has lasted for over 30 years. The seed system wasn't designed to work this way, but it is working.

References

Cheema, S.S., Kolawole, K.B. and Oyebanji, O.O. (1994) Agricultural development projects and national seed service linkages. In: Adedipe, N.O., Usman, I.A. and Maini, N.S. (eds) *Evolving the Nigerian Seed Development Plan.* Federal Department of Agriculture, Abuja, pp. 39–49.

Joshua, A. (1994) Privatisation of the Nigerian seed industry: the way out. In: Adedipe, N.O., Usman, I.A. and Maini, N.S. (eds) *Evolving the Nigerian Seed Development Plan.* Federal Department of Agriculture, Abuja, pp. 25–38.

Madukwe, M.C., Okoli, E.C. and Eze, S.O. (2002) *Analysis and Comparison of the Agricultural Development Programme and University Agricultural Technology Transfer Systems in Nigeria.* ATPS Working Paper Series No. 35. African Technology Policy Studies Network, Nairobi.

Shobowale, O.J. (1994) Strategy for fulfilling foundation and certified seed requirements. In: Adedipe, N.O., Usman, I.A. and Maini, N.S. (eds) *Evolving the Nigerian Seed Development Plan.* Federal Department of Agriculture, Abuja, pp. 121–139.

Tripp, R. (2001) *Seed Provision and Agricultural Development: the Institutions of Rural Change.* Overseas Development Institute, London. 174 pp.

Usman, I.A. (1994) Welcome address. In: Adedipe, N.O., Usman, I.A. and Maini, N.S. (eds) *Evolving the Nigerian Seed Development Plan.* Federal Department of Agriculture, Abuja, pp. 1–3.

5 Mali: When Government Gives Entrepreneurs Room to Grow

DANIEL N. DALOHOUN, PAUL VAN MELE, EVA WELTZIEN, DIOUKAMADY DIALLO, HAMIDOU GUINDO AND KIRSTEN VOM BROCKE

5.1 Introduction

5.1.1 Agriculture

Mali is a vast country, landlocked in the heart of the West African Sahel, covering 1,241,238 km^2 with an estimated 13 million people. More than 65% of its land is desert or semi-desert. The country is fed by the great Niger and Senegal Rivers that shape Mali's young, market-based economy, still largely dominated by subsistence farming, herding and fishing. Industry is based on food processing, some textiles and gold and phosphate mining. The economy remains vulnerable to price fluctuations in its two main exports: gold and cotton.

Economic growth was hampered by the state-led strategy adopted at independence, which led to unwieldy parastatals clogging up key economic sectors, weak infrastructure, burdensome administration and poor social conditions, including a low literacy rate and a population growth of 3.3% per year (one of the highest in the world in 2009; John F May, personal communication). Mali still depends on foreign aid, as the government keeps implementing economic reforms and free market policies to meet the expectations of donors and private investors.

Mali covers three climatic zones: the desert-Saharan zone in the north with less than 200mm rainfall; the Sahelian zone in central Mali with 200 to 600mm rainfall; and the Sudanese zone in the south with 600mm to 1400mm of rain falling from late May to October.

Essential food crops such as maize, millet, sorghum and fonio are rainfed. To promote irrigated rice, large hydro schemes have been built in the inner Niger delta and in the valleys of the Niger and Senegal Rivers. Agriculture is dominated by small, family farms and the main way to harvest more food is simply to farm more land.

Cotton is no longer king. Upland rice has made its entry, and the inland valleys in southern Mali are increasingly used to grow rice in rotation with vegetables or potatoes.

5.1.2 Evolution of the seed sector

Informal sector. Farmers growing vegetables, maize and irrigated rice use more improved seed than those growing traditional crops. In the Koulikoro region (1000 mm of rainfall) less than 8% of seed of new sorghum varieties was introduced in villages through purchase (Siart, 2008). Most entered through exchange, gift or inheritance. Once a new variety is adopted, farmers produce their own seed as much as possible.

However, it is a stereotype to say that farmers buy no sorghum or pearl millet seed. Farmers do buy grains in the market, when they spot interesting traits or pure varieties, and then plant them (Siart, 2008).

Sorghum and millet seed is sold at weekly fairs in the districts of San (450–600mm of rain) and Douentza (receiving only 200–400mm), and, while it is not certified (Smale *et al.* 2008), the vendors are mainly women farmers who bring grain that is suitable for seed directly from their granaries to the market. Seed of local varieties is neither packaged nor labelled but identified by its provenance. However, it is never sold explicitly as 'seed' because customarily *la semence ne se vend pas* ('seed is not sold'). As buying seed makes one a bad farmer, local markets can be a means of impersonal exchanges without social stigma (Smale *et al.*, 2008).

Under harsh and variable conditions, local sorghum and millet varieties often perform better than improved varieties. They can adjust the length of their growth cycle to synchronize with the length of the rainy season (called photoperiod sensitivity). Unfortunately, early breeding programmes, combined with the effects of drought, gradually eliminated photoperiod sensitivity in favour of varieties with fixed, short cycles. These shortcomings have since been overcome (Weltzien *et al.*, 2007).

The formal sector has undergone many changes since 1996 when state-owned seed production began to be transferred to farmers' organizations and other private producers as part of Mali's broader market liberalization strategy (Warms, 1994;

Dembélé *et al.*, 2003). Operation Production of Improved Seeds (OPSS – Opération Production Semences Sélectionnées) was created by the government in 1977 to produce, collect, store and distribute seeds. In 1991 the OPSS became the National Seed Service and worked closely with a network of farmers on state farms to produce certified seed.

To support the National Seed Service as the state disengaged, the Seed Sector Support Project (PAFISEM – Projet d'appui à la filière semencière) started in 2003 with support from the African Development Bank (AfDB) to: (i) create a network of seed producers in each region; (ii) strengthen the technical and financial capacity of seed producers; (iii) equip and strengthen the Seed Laboratory (LABOSEM – Laboratoire de Semences) and decentralize its activities; (iv) support actions to create infrastructure for storing certified seed; and (v) establish a National Seed Fund (for national security stock and to be prepared for disasters).

PAFISEM supported 136 cooperatives and various private enterprises, such as Faso Kaba (Section 5.2), Comptoir 2000, Nakoshi and Niégué Farm (Section 5.3). After the project came to an end in 2009, the various organizations approached the Malian government to continue subsidizing seed certification, at least for the costly field inspections. While the project supported the transition towards a private seed sector, it focused on seed production, but hardly at all on business management, seed distribution and marketing. The Nerica dissemination project of the African Rice Initiative (ARI) started addressing this for rice seed producers' organizations, e.g. by establishing revolving funds (see Chapter 3 for a similar experience in Cameroon).

The Seed Association of Mali (ASSEMA – Association Semencière du Mali) was created in 2003 under the umbrella of the African Seed Trade Association (AFSTA). It started with seven traders, focusing on imported vegetable seeds and agrochemicals. In 2009 it had 20 members and a network with 26 government agencies, 12 national non-public organizations and 25 international organizations.

5.1.3 Government support for the financial sector

The National Bank of Agricultural Development (BNDA – Banque Nationale de Développement Agricole) was created in 1972. Farmer cooperatives, groups or associations could apply for loans for agricultural equipment, tools and agrochemicals. Later on, it started financing the agro-industry, especially cotton, which accounted for about 70% of its activities. The crisis in the cotton sector coincided with the deregulation of the banking system in French-speaking West Africa, allowing the bank to finance activities in other sectors (Box 5.1). Today, 40% of the bank's investments are for rural financing, including seed production.

PAFISEM paid for seed certification and had a guarantee fund to reimburse banks for unpaid loans from seed producers. Farmers and their organizations did not know about this secret agreement; otherwise they would not have repaid the loans.

Seed producers need financing for a longer time than grain growers. The BNDA offers loans at relatively attractive annual interest rates of 8–10% for cooperatives and farmers' associations and 12–14% for private companies.

In addition to loans to buy equipment and other agricultural inputs, cooperatives and farmer organizations can also apply for credit to purchase seed from their members (which they process and store for sale during the next season). The amount of credit BNDA grants to seed producers depends on the number of members of the cooperative, how much land they have, the crops produced, the economic situation of its members and the type of financing sought.

Box 5.1 Termites speed up rural financing.

In the mid-1980s, the Malian government realized that farmers kept their savings from cotton sales under couches, in the horns of animals or in holes. However, termites often ate the money before it could even be invested.

As the government was opening up to the market economy, it wanted to stop the Compagnie Malienne de Textile (CMDT), a state-owned enterprise working with cotton farmers, from managing credit. Supported by the international community, the government approved the creation of a union of savings and credit establishments.

The Ministry of Finance established a regulatory framework for the central bank to supervise the management of these union funds, which later became known as Kafo Jiginew, the largest micro-finance institution in the country. In 2009, it had over 160 branches and 250,000 members, of which 65% are farmers. Seed producers in particular are known to be faithful in repaying their loans.

Now that farmers have a safe place to save, there is money to loan, but not for termites.

In 2010, as PAFISEM nears its end, the Ministry of Trade intends to provide a guarantee fund to BNDA and the Banque Malienne de Solidarité (BMS), among other banks, to support the emerging private sector. The main objectives are to lower the interest rates and to extend the loan periods. Both the guarantee fund and the subsidized interest rates need to be implemented with caution to ensure that farmers repay their loans to the banks (World Bank, 2008). Uganda is also strengthening its agriculture through the financial sector (Section 10.5).

5.1.4 Agricultural law and seed policy

A new agricultural development framework law (LOA – Loi d'orientation agricole) was enacted in 2006 to promote sustainable, modern and competitive agriculture, based primarily on family farms. It stresses the production, marketing, diffusion and adoption of certified seed and encourages private investment.

The latest draft of the national seed law explicitly allows only for certification for seed produced from varieties registered in the seed catalogue (Diakité *et al.*, 2008). Yet certified seed production is still low (Table 5.1).

Table 5.1. Seed certified (tonnes) in Mali, 2005–2009. Source: LABOSEM.

	2005	2006	2007	2008	2009
Maize OPV	177	1,725	764	860	865
Rice	953	2,258	1,298	1,121	1,454
Sorghum	43	95	122	436	510
Millet	43	123	95	162	436
Groundnut	7	14	7	4	23
Cowpea	9	10	8	17	18
Total	1,232	4,223	2,294	2,599	3,306

The Seed Laboratory under the Ministry of Agriculture is in charge of all seed certification activities. It is the technical office of the Direction National de l'Agriculture (DNA), which deals with regulatory seed control and certification. The laboratory is accredited by ISTA (International Seed Testing Association) and standards are taken from the OECD (Organisation for Economic Co-operation and Development) to encourage the movement of seed within ECOWAS (Economic Community of West African States). Technicians give advice to seed growers and follow seed production from soil preparation, to sowing, harvesting and processing.

5.1.5 Seed demand and use

Mali released four hybrid sorghum varieties in 2008 and 2009 and plans to release five hybrid maize varieties. Demand for seed of improved varieties is generally perceived as being low. However, experiences show that, when superior varieties are available and farmers know about the specific advantages and the availability of the seed, then demand grows (Box 5.2), especially when new opportunities for grain marketing evolve.

Demand for seed of improved varieties is increasing, and is higher for irrigated than for rain-fed rice. Farmers grow most of their own seed for staple food crops, sorghum and millet, or they exchange within their village (Siart, 2008). Demand for seed of a new variety may be low initially, and growing demand can often be met within the village. Improved sorghum varieties have been more adopted than those of pearl millet (Smale *et al.*, 2008).

Farmers require sorghum and pearl millet varieties that are adapted to their climate, soils, pests, diseases and other constraints. Farmers plant various varieties, to match the growing conditions of each field. New varieties are usually not evaluated as potential replacements of existing ones, but as additions to the village's variety portfolio (Box 5.2). In pearl millet, due to its outcrossing nature, farmers may actually be preserving multiple characteristics within an identified variety.

> **Box 5.2 Demands for seed diversity.**
>
> When breeders included an extra early maturing sorghum variety, Diacumbe, in trials for the Sudanese zone, many farmers reacted angrily, as birds were eating the grains before they could be harvested. A few years later one of the same farmers requested foundation seed of Diacumbe for seed production. Demand for this 'unadapted' variety was rising because some farmers started using it in fields near their houses, where bird scaring is possible. They could thus harvest before anyone else, fetch very high prices and address the food gap in the 'hungry season', just before general harvest. Other farmers started using it for very late sowing, when all other varieties would fail, because they would be hit by end of season drought before the grains could fill.

Farmers often buy legume seeds in markets because of pest damage in storage. Local adaptation is less important for legumes; they can be grown in a much wider range of growing conditions than cereals, which need to mature after the rains have ended to obtain good grain quality.

The diverse seed enterprises described below illustrate the recent developments in the formal seed sector. The first one is about a dynamic woman who established Faso Kaba, one of the few seed trading enterprises. The second case is Niégué Farm. When state-owned farms were privatized, a retired agronomist in the Office du Niger changed it into a thriving rice seed enterprise. The third case describes the Nipagnon Cooperative in the Sikasso region, producing seed of many food security crops. The Mandé Seed Cooperative in Siby, about 50 km from Bamako, collaborated with ICRISAT to select promising sorghum lines until they became sorghum seed producers. The last case describes one of the Dogon villages north of Mopti that produces quality seed and is increasingly recognized for its early maturing pearl millet.

5.2 Faso Kaba

5.2.1 History

Faso Kaba Sarl (Ltd) is a seed company registered in 2007 in Bamako, Mali, that started as a seed dealer in 2005. However, the story goes back to the 1980s.

Best use of leisure time. The owner of Faso Kaba, Mrs Coulibaly eagerly shared her story. 'In 1985, my husband obtained a PhD scholarship in maize breeding in the United States of America. I went with him, but did not like to just sit at home every day while my husband was studying. I needed a job to keep myself busy with something useful.' Soon after, she obtained a position as a labourer in the Garst Seed Company, an early innovator of hybrid maize seed in the USA widely recognized for its unique production and distribution methods to deliver high yielding maize, soybean, sorghum, lucerne and sunflower seed.

Working at the maize unit, Mrs Coulibaly was in particular intrigued by the art of delivering high quality seed to farmers. 'But each time I thought of the conditions under which Malian farmers worked and the meagre results they got a sadness overwhelmed me. Rather than becoming frustrated, the challenge of providing seed related services (in Mali) came to my mind. Once back home, I was eager to achieve this dream.'

From informal to official. When Mrs Coulibaly returned with her husband she had to start a new life in Mali. Helping her husband breed new maize varieties had become her daily passion. The quality and yields of those varieties were impressive and attracted many producers. Men and women, sometimes in groups, approached her for seeds. But people did not know that her husband used foundation seed. Mrs Coulibaly persuaded her husband to multiply the foundation seed, which he did on a small plot, just about 0.25 hectare. Under his supervision, Mrs Coulibaly took care of the field and harvested it. She bought bags and packed the seeds for sale. The success of this experience encouraged her to expand.

Mrs Coulibaly noticed that the seed-producing groups and cooperatives supported by the PAFISEM project were isolated from potential customers. With few resources she started collecting seeds, packaged and stored them for sale during the next season. To face the fast growing demand for maize seed she registered as a seed retailer in 2005 and opened her shop in Bamako.

From a registered retailer to a company. As she increasingly supported seed producers, in 2007, Mrs Coulibaly transformed her retail business and officially registered as Faso Kaba. In the Bambara language, Faso Kaba means 'maize of our land'. The company did a market study to identify the needs of seed producers and assess cereal growers' evolving interest in quality seed, and also of other crops. ICRISAT, the National Seed Service and other organizations encouraged Faso Kaba by providing information.

Expanding to other crops and products. Family farmers produce many crops and their experiences with maize led them to seek good seed for other crops. To respond to their demand, Faso Kaba branched out, trading in certified seed of rice, cowpea, sorghum, millet, groundnut, gombo (okra), potato and sesame (Table 5.2).

Table 5.2. Seed sold (tonnes), Faso Kaba.

	2005	2006	2007	2008	2009
Maize	1	20	41	60	180
Rice	–	–	8	70	42
Cowpea	–	–	0.2	9.5	15
Sorghum	–	–	0.5	5	15
Millet	–	–	0.2	3	10
Okra	–	–	0.1	0.2	0.5
Groundnut	–	–	–	10.5	20
Potato	–	–	–	2	2.5
Sesame	–	–	–	1.3	3
Total	1	20	50	162	288

Potato seed is imported from Europe via the ports of Senegal (Elodie variety) or Abidjan (varieties Claustar, Appolo, Spunta and Sahel), and transported by truck or train to Bamako. Faso Kaba also started selling fertilizers, pesticides and sprayers, which its marketing study had revealed as complementary products to sell.

The Alliance for a Green Revolution in Africa (AGRA) did not believe that a company can master seed distribution and quality if it does not also produce seed, so in 2007–2008 it supported Faso Kaba in seed production and marketing. The company started growing its own seed of maize (2007), later on adding sorghum (2008), groundnut and millet (2009). To supplement its own production, Faso Kaba continues to tap into its network of seed producers to meet the demand.

'But rice, cowpea and groundnut seed production is a bit complicated,' said Mrs Coulibaly. 'Farmers need more training in the production and processing of these seeds to meet certification standards. They are more experienced with the production of certified seed of maize, sorghum and millet.'

5.2.2 Structure

Management and staff. Mrs Coulibaly is the sole proprietor. She is the director, manager and accountant of the company. She employs a driver, two labourers and a seed specialist who evaluates the quality of seed to buy and advises on storing and stock-keeping. The business grew rapidly and steadily and in 2008 Faso Kaba hired an agronomist to look after the farm, assisted by a technician with 20 years of seed production experience in Mali. Finally, Mrs Coulibaly hired a stock keeper and an accountant when she could no longer manage all the tasks

Seed entrepreneurs need to strengthen their networks and invest in trust building. Mrs Coulibaly from Faso Kaba does that on a daily basis.

by herself, even the ones she could do well. The company holds weekly staff meetings to brainstorm on solutions. In June 2010 Mrs Coulibaly received the prestigious African Business Award for Best Agricultural Development.

Infrastructure and equipment. Faso Kaba produces some seed itself, but continues to collect, transport, package, store and sell seed of multiple crops. It has office equipment and rents its office and shop, along with the two adjacent stores that can stock 40 tonnes of seed. Faso Kaba has its Bamako shop at a strategic location along the road to Ségou where anybody travelling upcountry has to pass. This is especially attractive to the Sunday farmers (people with full-time jobs who spend free time working the land and often send seed to their parents in the village). In Banankoroni, about 5 km from its Bamako shop, Faso Kaba is currently expanding its warehouse on a 0.5 hectare plot to store up to 500 tonnes of seed.

Faso Kaba relies on public transport (taxi, minibus) or small trucks to haul seed from the seed producers to Bamako for packaging, after which it distributes seed to its 50 selling points using similar means. Faso Kaba pays 13 FCFA ($0.03) per kg seed transported.

Seed producers. During the marketing study Mrs Coulibaly learned about government and private maize and rice seed producers, even in remote areas. Apart from some individuals, they were mostly cooperatives composed of former farmer-employees of state-owned farms that had been privatized. Faso Kaba partially obtains seed by establishing seed production contracts, advancing foundation seed (90% coming from the Institut d'Economie Rurale – IER), fertilizers and agrochemicals and buying back the certified seed while deducting the costs of the loan. The price paid for the certified seed is determined only at the time of harvest, as it depends on the market price.

Most seed producers do not necessarily depend on Faso Kaba. They produce seed without contracts and sell to various clients. Faso Kaba can contact any of the seed-producing cooperatives or associations if it receives orders that are not covered by any of its contractual seed producers.

Seed dealers. To establish seed selling points, Faso Kaba organized community meetings to help identify contact people in the regions of Koulikoro, Ségou and Kayes. Often agro-dealers or farmers linked to a community-based organization were proposed. Nearly all were farmers themselves. In 2007 Faso Kaba worked with 33 seed dealers, and by 2009 this had grown to over 50, of which ten are women. One of Faso Kaba's local dealers in Dialakoroba village sells seed to over 30 neighbouring villages. All were trained by IER and extension staff, as part of a wider effort of AGRA to strengthen agro-dealers in Mali. Most village shops are only open a few months per year at the beginning of the growing season. In Ségou, however, Faso Kaba shares rent and taxes with a shop owner to ensure it is open throughout the year.

Quality control. Faso Kaba only sells certified seed. Seed is inspected at the source by formal inspectors when it is packed in large bags of up to 100 kg and kept in the farmers' or cooperatives' stores. However, for laboratory tests they only take a 2 kg sample from each seed lot (which can be up to 20 tonnes if it is from a single variety grown in a single village). Hence, when the seed is hauled to Bamako, Faso Kaba mobilizes 25 labourers to clean the seed again before packaging in smaller bags as it cannot afford to have any immature seeds or debris in the seed it sells.

When seed is grown under contract, Faso Kaba pays for the inputs and the certification (laboratory tests only; the field visits have so far been covered by the government). But more and more Faso Kaba buys directly from seed producers who covered these expenses themselves.

Land. To respond to AGRA's demand to get involved in seed production, Mrs Coulibaly requested and received 60 hectares of land from her husband's clan. By 2009 she grew maize, sorghum, millet and groundnut seed on 20 hectares.

Quality control and customer-tailored packaging are some of the key strategies to build reputation.

5.2.3 Cash flow

Banks were ready to grant credit to buy agrochemicals, but they did not think that buying and selling seed were profitable. So Mrs Coulibaly set up Faso Kaba with her own money. Despite the banks' lack of imagination, Faso Kaba grew, but it needed more cash. With one rain-fed cropping season cereal seed is sold for only 3 months, so selling vegetable seed and fertilizers helps to spread income over the year. 'To be able to respond to opportunities quickly, bank loans should be for at least 2 years,' explains Mrs Coulibaly. 'If you have taken a loan in April and you need to pay back in September, you have no money in October to buy seed potato, which farmers want after the rainy season, so you have to start a new dossier for credit all over again.'

In 2007, AGRA decided to strengthen Faso Kaba's distribution and marketing capacities and help it to buy seed from growers. In practice, AGRA persuaded the growers to sell their seed on credit to Faso Kaba. AGRA guaranteed farmers that Faso Kaba would pay them at the due time. As the project came to an end, Faso Kaba hopes that the government will act quickly to make loans more business-friendly.

For big seed orders, Faso Kaba asks its clients to pay 60% advance and the remaining 40% upon delivery. When selling to her network of 50 local seed dealers she tries to limit selling on credit as much as possible, but there are no fixed rules. Mrs Coulibaly believes the seed sector would benefit from a specific credit line for agro-dealers.

5.2.4 Marketing

Initially, to solicit maize growers' demand, Faso Kaba bought three mobile phones and gave one each to three farmer representatives in three village clusters, which it wanted to set up as local seed dealers. The representatives shared their mobile number with other farmers through their local radio. Farmers soon started placing seed orders, which the local dealers collected and passed on to Faso Kaba by phone and in writing with a taxi driver to Bamako. By the time the driver arrived, Faso Kaba had already prepared the order and handed it over to the taxi driver who returned to the village with the seed. Faso Kaba continues to use this method and has expanded its distribution network to 50 villages across three regions, several hundreds of kilometres apart. In 2010 the network will be expanded to the Sikasso region.

All local dealers are motivated to sell. They receive 25 FCFA ($0.05) per kg of cereal seed sold, 100–500 FCFA ($0.22–$1.10) per bag of vegetable seed and 250–500 FCFA ($0.55–$1.10) per bag of fertilizer sold, the amount depending on the volume.

In 2009, the average price per kg certified seed sold was 350 FCFA ($0.77) for maize, sorghum and millet; 500 FCFA ($1.10) for rice; 750 FCFA ($1.65) for cowpea; and 900 FCFA ($2.0) for groundnut.

Faso Kaba's main clients nowadays include Sunday farmers, the 50 local dealers, full-time farmers, cooperatives, NGOs and occasionally FAO and other projects. However, following its marketing campaign, increased visibility and reputation among farmers, by 2015 Mrs Coulibaly predicts that most of her seed will be sold to local dealers (Table 5.3). About 60% of Faso Kaba's clients are repeat customers who return to buy seed every year, irrespective of the crop.

The company's forte is its marketing strategy, which it strengthened in 2008–2009 with AGRA support. First of all a radio programme was made that stressed the advantages of improved seeds. Faso Kaba made 50 copies and gave one to each of its local dealers, who in turn took it to their local radio station. These broadcast the programme four times per month prior to the cropping season, for each broadcast receiving 1000 FCFA ($2.2). The radio broadcasters frequently invited the local seed dealers during call-in sessions, during which they announced that seed is available.

Table 5.3. Clients of Faso Kaba.

	2005	2009	2015 (predicted)
Sunday farmers	1	1	4
Local dealers	–	2	1
Individual farmers	–	3	2
Groups and cooperatives	–	4	3
Projects and NGOs	2	5	6
Government	–	6	5

Ranking assessment by senior management of seed enterprise, 1 being the most important.

Faso Kaba also advertises directly through the national media, e.g. on Friday on the agricultural programme 'Le Monde Rural' put on by the national radio ORTM. When Faso Kaba attends seed fairs or organizes field days it often invites the national TV, which will comment on them 5 to 15 minutes after the nine o'clock evening news against payment, or on Monday at 6 p.m. when farmers are home. Broadcasts are done in either French or Bambara (one of the national languages). 'Increased visibility pays, as I receive telephone calls from new customers every day,' says Mrs Coulibaly.

Faso Kaba earned customers' trust by selling them quality seeds and services. Faso Kaba occasionally sells agrochemicals to farmers on credit and its agronomist gives technical assistance to farmers to ensure good production. At harvest time, farmers pay back their loans without having to pay interest. It is difficult to win farmers' trust, and yet so easy to lose it (Box 5.3).

Faso Kaba also puts an entry in the yellow pages, and hands out business cards, calendars, caps and T-shirts during grain fairs and cereal fairs, organized by NGOs or the Ministry of Agriculture.

Another important marketing strategy is the packaging. Depending on the demand, cereal seed is sold in bags of 0.5 to 40 kg, vegetable seed in bags of 5 to 500 grams and seed potato in 25 kg bags. If farmers or farmer groups are reluctant or doubtful, Faso Kaba hands out small packs of seed between 100 g and 3 kg for farmers to try them out. All bags contain technical information on the seed and the variety.

The company now sees that opportunities for export are increasing as the Mauritanian government is emphasizing certified seeds. Besides Mauritania, the company also exports seed to Burkina Faso, Chad, Niger and Senegal based on orders placed by visitors at the annual seed fairs organized by the Ministry of Agriculture.

Box 5.3 A bad experience turning sour.

The importance of quality and certified seed was highlighted by a regular customer we happened to meet at Faso Kaba, in Bamako.

In 2007, he collected a seed order from his colleague farmers, travelled to Bamako, and figured he could make more money buying seed *tout venant* (whatever comes my way) from the market. This impure seed was cheaper than the seed he used to buy for them from Faso Kaba.

'I bought maize seed *tout venant* and distributed it to others in my village. Because of the bad quality of the seeds, yields were meagre that year and the farmers refused to pay for the seeds that I sold them on credit. I knew it was my fault and I lost about 10 million FCFA ($2200). From that day on, I swore to go only for certified seed.'

5.3 Niégué Farm

5.3.1 History

When the government privatized all state-owned farms in 1994, Ousséini Doumbia, a retired agronomist, won the bid on one of these farms in the village of Niégué, in the Office du Niger (a state agency created in 1932 to manage a more than 70,000 hectare irrigation scheme in the interior delta of the Niger River). Together with nine colleagues Mr Doumbia started managing the farm. Five years later he and a few other elders left management to become just members.

The (partially) new management team that took over in 1999 is still in place. In 2000 they registered themselves as the Association of Economic Interest for the Rational Exploitation of Niégué Farm. According to the current head, Agouno Ongoiba, 'moving into the seed business was the best choice I ever made'. Mr Ongoiba is also an agronomist and former director of one of the five zones of the Office du Niger. Rice seed production on a developed land with irrigation facilities is a lucrative business.

Being a rice seed specialist with a long experience and the right know-how helps to guarantee high quality. And there is a growing market for seed.

Due to fertilizer, training the workers and better organization average yields on the Niégué Farm increased from 5.2 tonnes per hectare in 2007 to 5.8 in 2009. The average waste (after processing into seed) decreased from 16.7% in 2007 to 13.0% in 2009. Only rice seed is produced (Table 5.4).

5.3.2 Structure

Table 5.4. Rice seed produced (tonnes), Niégué Farm.

	2005	2006	2007	2008	2009
Kogoni 91–1	66	82	75	90	93
Adny 11	18	23	42	36	58
AD 9246	20	23	29	33	32
BG 90-2	20	21	14	29	26
Wat 310	1	1	12	24	37
ECIA	8	10	13	3	3
RPKN2	3	3	1	0.6	1
Séberang Mh 77	4	6	2	3	–
Total	148	180	195	223	260

Management and staff. The Niégué Farm is technically sound. Its staff includes four retired agronomists, two agricultural technicians, two economists and an accountant. The agronomists were extension agents who worked for many years with farmers, with whom they still have excellent relations. The technicians regularly visit their clients and help them. The stock keeper helps the supervisor manage the seeds and agrochemicals in storage. The farm occasionally hires groups of young labourers for farm and postharvest tasks. Over the years they have built a particular expertise in operating the association's equipment. However, these groups are becoming rare, more demanding and expensive.

Land, infrastructure and equipment. The Niégué Farm has 45 hectares in ten widely spaced pieces of land. When the farm was privatized it had some infrastructure and equipment. On 4 hectares there is a functioning irrigation and drainage system that allows off-season production and irrigation in the rainy season when the rains fail. To use this system the farm pays an annual fee of 3 million FCFA ($6500) to the Office du Niger, which owns it. Two big stores with a capacity of 200 and 300 tonnes of seed were also part of the deal. There is one generator and one seed processing system for seed preparation, cleaning, drying, coating, packaging and storing.

Production of quality seed. Each member of Niégué Farm is assigned one of the ten fields and produces only one of the eight popular rice varieties. Nobody is allowed to change the variety he produces. The objective is to ensure high quality seed. Producing the same variety year after year avoids seed mixtures and provides greater experience with the variety, both of which guarantee quality seed. To avoid volunteers (rice plants that sprout from the grain spilled in the field), the seed growers water the soil some time before the season to make these seed germinate and later plough them under. After harvest cattle graze on the stubble. The farm cannot increase production since no more land is available. If demand exceeds what the farm can produce, Niégué Farm would need to contract outgrowers.

Links and partnerships. There is a strong relationship between the Niégué Farm and farmers, who are the bulk of its customers. Niégué Farm is also tied to IER, which supplies it with foundation seed. The association estimates the demand for seed of different varieties based on recent demand or the expectations of farmers and channels this information to the agricultural research institute. The farm has never benefited from any technical or financial support from NGOs, projects or any other organization.

The association is run like a private business, pure and simple. It has no particular relations with competitors except when they meet for training or information that requires gathering seed producers.

5.3.3 Cash flow

Prices of inputs such as agrochemicals and labour are increasing every year. If seed prices were to increase, farmers might be discouraged from buying seed. This price squeeze reduces the profit margin of seed growers, who face the same increase in production costs as grain growers. Because the Niégué Farm is an association, not a cooperative, it pays the full amount for seed inspections and seed certification services. Despite this, producing rice seed is still profitable according to its managers.

Niégué Farm is also successful thanks to loans from BNDA, which offers an attractive annual interest rate of 12%. Conditions are negotiated and fixed in contracts with the bank. So far the farm has always repaid the loans promptly.

Except for some rare situations, customers do not need credit from Niégué Farm. Only some seed dealers require a few days to pay for their purchase.

It is unlikely that new rice varieties will easily compete with the ones from IER, so Niégué Farm does not fear immediate competition. Farmers need to get to know a variety before they will buy it. And Niégué Farm has always sold all its seed, thanks to its quality and the follow-up services provided by the agricultural technicians.

5.3.4 Marketing

Seeds produced by Niégué Farm are sold only in Mali. Individual farmers buy about 80% of the total production. Many of them have asked to represent the farm to distribute seeds in their villages. Other clients include the FAO, NGOs and projects. Prices have changed little since 2003. From 250 FCFA ($0.55) per kg in 2003, a kilogram of rice seed was sold at 300 and 330 FCFA ($0.65 and $0.73) in 2006 and 2009.

The farm started marketing by organizing field days announced on local radio to explain that, under the right conditions, certified, quality seed allows higher yields and better grain quality. It creates higher market value and can increase farmers' incomes. All this was shown on

When farmers hear on the radio that there is a field day or a seed fair in their area, they are delighted to attend.

demonstration plots and visitors could see the behaviour of different kinds of seed. Rice seed is packaged in 50 kg bags printed with the name 'Niégué Farm.' Thanks to its permanent contact with research, the farm is aware of any new varieties, and tries them before recommending them to customers.

5.4 Nipagnon Cooperative

5.4.1 History

After the state-owned seed farms collapsed or were privatized, people were encouraged to create cooperatives or other organizations to take over seed production. One of these was the Nipagnon Cooperative, created in 2006 in the region of Sikasso, southern Mali. The cooperative has 35 members, including six women. It sells seed produced by its members.

PAFISEM helped build a storage building, trained producers and regularly controls seed production for certification. The international NGO Sasakawa Global 2000 (SG2000) contributed to 40% of the cost of the administration building. In contrast to the Niégué Farm described in Section 5.3, more than 90% of the cooperative members are illiterate. The cooperative does not own collective land except for 1 hectare used to grow okra.

The cooperative produces seed of maize (Debagnonman and Sotubaka varieties), rice (Nerica 4, Nerica 8, Nerica 9 and Nerica 12), sorghum (CSM 388), millet (Djiguifa) and soybean (G196) (Table 5.5).

Table 5.5. Seed produced (tonnes), Nipagnon.

	2006	2007	2008	2009
Maize	50	55	62	10
Rice	28	30	32	44
Sorghum	2.0	2.2	3.1	3.8
Millet	1.1	2.1	3.0	3.2
Soybean	0.5	0.8	0.8	1.8
Total	82	90	101	63

5.4.2 Structure

Management and staff. Fifteen members founded the cooperative, and others kept joining later. Three members left in 2008: they did not respect production norms and the certification services rejected their seed. The profitability of seed increased the seed producers' incomes, attracting other farmers to join the cooperative in 2009.

Most members lack business management capacities. The president of the cooperative said, 'We are peasants and not intellectuals. We don't know how to manage. We should hire someone, but it would cost money the cooperative cannot afford.' Indeed, apart from the secretary and the president, a respected 56-year-old nobleman, none of the members can read or write. Currently, management has five elected members, with three men occupying the positions of president, vice-president and secretary. The treasurer and the agent in charge of information and organization are women.

There is a mutual trust among the members and an eagerness to perform better than other cooperatives involved in producing seed. They are also motivated by their ambition to build a viable and powerful cooperative, to be proud to be its members and to acquire a higher social status with increased income and investments.

Land. Land is abundant. The village offered 1 hectare of land to the Nipagnon cooperative for building and development.

Infrastructure and equipment. Most farmers work manually, but use fertilizers and pesticides. There is no irrigation service as in the Office du Niger and farms are rain-fed. The cooperative owns a storage building, a machine room, a complete seed processing chain and some tools. It acquired this equipment thanks to hard work, good management and some valuable donations.

In their early days most cooperatives receive infrastructure or equipment through outside support.

5.4.3 Cash flow

The Nipagnon Cooperative uses its profits to pay for storage pesticides and the foundation seed that it sells on credit to its members.

However, financial worries start when it is time to buy back the certified seed from the members. Then Nipagnon obtains loans from BNDA or a rural credit organization called Kafo Jiginew (Jiginew is the Bambara word for the traditionally woven purse that older men tie around their waist). The cooperative can take a loan between 500,000 and 2 million FCFA ($1000–$4000). It repays the loans once the seed is sold.

The cooperative also helps its members to manage their incomes. The members have decided to always keep part of their income with the cooperative to buy agrochemicals and to pay for labour. The needs and costs of these are estimated for each member, based on the area he or she aims to cultivate. The amount is deducted from the member's sales and the rest is paid to him or her. This way, the members get the inputs they need. After the initial financial, technical and marketing support from SG2000 newcomers to the cooperative received training from PAFISEM.

5.4.4 Marketing

Nipagnon Cooperative's major customers include two seed trading companies, Faso Kaba (Section 5.2) and Comptoir 2000, individual farmers, projects and various NGOs. It does not anticipate any changes over the next years.

Nipagnon also advertises over local radio stations in Bambara, the local language. The programmes were carefully prepared with the support of SG2000 and PAFISEM to explain that certified or good quality seeds improve yields and grain quality so farmers can earn more money. The ads stress the unique quality of Nipagnon seeds, produced under the guidance of seed specialists and certified by PAFISEM.

By attending the annual seed fairs organized by the Ministry of Agriculture, the cooperative started exporting seed to Burkina Faso, Guinea, Mauritania and Senegal. About 40% of the production is sold to Senegal.

5.5 COPROSEM

5.5.1 History

Officially registered in 2006 with headquarters in Siby, a village about 50 km from Bamako, COPROSEM (Coopérative pour la Promotion de la Filière Semence de Mandé) started with nine farmers, each from another village around Siby, who assigned administrative roles among themselves in a verbal process.

Before the cooperative started, a farmers' group helped ICRISAT (International Crops Research Institute for the Semi-Arid Tropics) researchers to identify the best sorghum varieties for their farming conditions, using a two-step participatory varietal selection (PVS) approach (Weltzien *et al.*, 2006). Some of the farmers, impressed by the new sorghum varieties, sought to get them in large quantities. They went to ICRISAT and asked for the seeds they had helped to select. But there was only foundation seed.

ICRISAT told them that this foundation seed needed to be multiplied into another generation of seed before they could give it to farmers. The farmers, all illiterate, organized themselves as COPROSEM to transform the foundation seed into certified seed. One of the first activities of the new cooperative was to participate in training sessions on procedures of formal seed production organized by IER and ICRISAT.

Table 5.6. Seed produced by COPROSEM.

	2006	2007	2008	2009
	Cultivated area (hectares)			
Sorghum	9.3	9.3	5.7	5.0
Maize	–	–	–	6.0
	Net production – certified and sold (tonnes)			
Sorghum	2.0	2.9	1.7	1.6
Maize	–	–	–	2.6

The farmers organized a seed fair that they advertised on local radio; ICRISAT paid the bill. Given its success, they launched a second fair and invited more farmers to learn about the advantages of good seed. The success made them realize that they could sell more.

The idea of a cooperative thus emerged after researchers suggested that farmers should sell quality seed of preferred varieties. Being a cooperative would also make it possible to acquire additional funding to remunerate seed producers in advance. Later, COPROSEM also produced maize seed (Table 5.6), and small quantities of uncertified seed of pearl millet, soybean and rice, to sell locally.

5.5.2 Structure

Management. Today, COPROSEM has 13 members with three others waiting to join. The youngest member is 35 years old and seven of the members are over 50 years.

COPROSEM members have acquired good seed production skills by engaging for various years in PVS trials with ICRISAT, but they lacked management skills. As with all cooperatives, the government assigned an extension officer to help them with production, sales management and record keeping. Additional management support was sought from the NGO Association Conseil pour le Développement (ACOD). The cooperative now keeps production and sales records.

COPROSEM's management team is composed of: (i) an administrative body which includes the president, the accountant and four secretaries for administration, supplies, information and commercialization; and (ii) a surveillance committee. A secretary in charge of seed stock management was added after the first experiences of seed production and commercialization.

Organization. Every year each member allocates some of his own land for seed production, growing household food and cash crops on the rest. The seed plot has to be isolated from other fields of the same crop. A few members use animal traction, but other farmers do all their work by hand. There is no irrigation.

At the beginning of each growing season, the members gather to estimate demands and talk about how to produce seed successfully. These meetings serve as training sessions for newcomers and as refresher courses for existing members. Seed inspectors start to make farm visits and continue through harvest and postharvest to seed certification.

5.5.3 Cash flow

Initially the costs for constructing a seed store, the labelling and the certification were covered by projects in collaboration with IER and ICRISAT, which also supplies them with sorghum foundation seed, at no cost. The cooperative knows it will have to pay for this in the future, as they also did for maize.

COPROSEM members plough, sow, weed and harvest their fields themselves so they do not need to pay for labour. As COPROSEM sells part of its certified seed to Faso Kaba, it gets fertilizer and pesticides on credit and shares them with its members.

As with most farm cooperatives, they need cash to buy the product (in this case seed) from their members and to process and store it for later sales. This is a difficult challenge which can make or break any cooperative. If a cooperative does not pay early, the seed producers will eat the crop as most are not food secure. In Mali credit lines are available for farmers and their organizations, but COPROSEM never bothered to apply. Rather it uses money saved by the cooperative from previous activities. As this money is not enough to buy and store larger amounts of seed, it limits the cooperative's growth. The extension officer assigned to COPROSEM is currently negotiating with the national farmers' association (AOPP – Association des Organisations Professionelles Paysannes) for them to hire an external accountant to prepare COPROSEM's balance sheets, which would qualify them to apply for credit from the BNDA or the Kafo Jiginew.

5.5.4 Marketing

The main customers include two seed distribution companies (Faso Kaba and Comptoir 2000), individual farmers, and other projects and programmes (Table 5.7). While companies buy most (80%) of the maize seed, individual farmers buy the bulk (75%) of sorghum seed. However,

Table 5.7. Clients of COPROSEM.

	2006	2009	2015 (predicted)
Seed companies	2	1	1
Projects and NGOs	–	2	2
Sunday farmers	–	3	–
Individual farmers	1	4	3
Members	–	5	4

Ranking assessment by senior management of seed enterprise, 1 being the most important.

this is likely to change over time as demand for quality seed is on the rise. In March 2010, Faso Kaba bought 3 tonnes of sorghum seed from COPROSEM at 300 FCFA ($0.65) per kg. This is slightly higher than what the cooperative sells it for normally, indicating that there is a real demand.

One of the members of COPROSEM is a trader. He started treating sorghum seed and putting it in 1 kg bags, which he sold at the local market at 400 FCFA ($0.88) per kg. At first it raised eyebrows among his colleagues, but now he has even started experimenting with 100 g seed bags, which he sells at 100 FCFA ($0.22).

Farmers are not yet aware that the cooperative has started producing maize seed. COPROSEM may need to return to its initial successful strategy of combining radio adverts with local seed fairs.

5.6 Pearl Millet Seed-producing Villages in Dogon Country

5.6.1 History and context

In the harsh Sahelian climate with erratic annual rainfall of 200 to 400 mm, pearl millet is the only cereal able to grow on the sandy soils. In this transition zone between the Sahara desert and the Seno plain of Dogon country in central Mali, pearl millet is the staple food for the Dogon people. Ever since the Dogon settled on the cliffs or bases of the escarpments in the 14th century, they have developed immense knowledge, cultural values, rules and powers associated with pearl millet.

Pearl millet was domesticated in this part of West Africa; its weedy relatives (locally called *chibra*) grow wild in the area, although they have bristles, shatter easily and mature before the cultivated form. As pearl millet is a highly cross-pollinated crop, with very good pollen longevity, coexistence with these wild forms may enhance its genetic diversity.

Dogon women collect the earlier-maturing weedy forms before the actual pearl millet harvest.

During the hungry period, women collect panicles of the weedy *chibra* before the pearl millet harvest. They carefully remove all the bristles before pounding the grains. Although both cultivated and wild forms can be used to prepare the staple food *toh* (pearl millet flour mixed with water heated above a fire and stirred into a paste, like mashed potatoes), *chibra* is unsuitable for couscous and crème (a local drink on the basis of water, pearl millet flour and tamarind that is consumed daily) and hence is always stored separately.

Farmers in the Douentza district, and especially in the north-eastern Haïre township, face some of the harshest conditions imaginable for crop production: temperatures at planting time can reach 50°C; the rainy season is short and erratic; and soils are sandy with a low pH (acidic). Any crop variety grown here must be adapted to these extreme conditions, be early maturing and able to resist noxious pests (such as headminers and the parasitic weed striga) and diseases (mainly downy mildew). Farmers in this area have developed clear rules and practices for managing seed, as their success in farming depends on having seed of well-adapted landraces.

Villages like Tabi, Tega and Toupere in Douentza district enjoy a good reputation in the Mopti region (central Mali) for the quality seed of their *Tabi nyo* (pearl millet from Tabi). According to farmers, their millet can mature in only 45 days, while it adapts its growth cycle to the length of the rainy season through continuous tillering from the base or from the lower nodes. It has thin, cylindrical, compact panicles 20 to 30 cm long, which are more resistant to headminers, an insect that feeds on the pedicels of the florets during grain filling.

Although farmers in Tabi were unsure about its origin, they knew of the uniqueness of their local landrace, both for growing and eating it. 'We know of farmers from other villages who planted our millet next to theirs and ours ripened first. Our millet is the best, because we have a good soil for it. At the market in Boni, women also prefer to buy two bowls of pearl millet from Tabi rather than four bowls of pearl millet coming from elsewhere. When they make *toh* from our millet, they become hungry less quickly.'

As pearl millet is highly cross-pollinating, characteristics of local landraces can change with time and farmers consider soil the most important influence on the development of plant types, similarly to pearl millet farmers in Rajasthan (Christinck, 2002). Farmers distinguish between the short millet types that grow well on hard soil (*aninam yu*, or 'pearl millet of the short people', referring to the Tellem, the pygmies who lived in caves and built dwellings around the base of the escarpment at the time when the Dogon arrived); another short pearl millet grown between rocks at the base of the escarpment (*torro yu*), but which can grow on either hard or sandy soil; and the tall pearl millet grown on the sandy plains (*dou yo*). Yields are highest for *dou yo*, but its larger, yellowish grains give a *toh* of lesser quality than *torro yu* or *aninam yu,* which have darker green, smaller, shinier and harder grains. *Torro yu* is preferred over the pearl millet grown on the sandy plains. When prepared, two bowls of *torro yu* give as much *toh* as three bowls of *dou yo* and *torro yu* is said to be more nutritious and heavy. Due to their different qualities in the field and in the kitchen, farmers often keep the different types separated, although they say that when one type is grown in the other environment it will become the other type after two seasons.

Many neighbouring villages have heard about the pearl millet from Tabi. Especially when rains are late, farmers from far away come to look for their very early maturing pearl millet. It has attracted customers from as far as Burkina Faso and Ségou (about 600 km south of Tabi). Recent trials with the variety in Maradi region, Niger, have excited farmers from more than 1000 kilometres away and have started seed production in Niger by ICRISAT.

5.6.2 Structure

Management. Farmers in Dogon country are not organized in farmers' associations or cooperatives, but abide by social rules and norms aimed at protecting the entire community. Although the amount of seed stocks are well-kept household secrets, by the time of planting farmers know who has no more seed or grain that can be used as seed.

Farmers in Tabi village also know that they should not buy pearl millet seed at the local market, and rather ought to ask people in their community when needed. But, as farmers are ashamed to admit that they have no more seed, many feel reluctant to ask others. Those who do will do so after dusk. Those who decline to ask out of shame will receive a (secret) visit the evening before the day on which all farmers go out to sow their fields. Some seed will be deposited in front of the house of the deprived neighbour, who may never know the origin of the donor. The importance of using seed from their own community is also reflected in village stories (Box 5.4).

Box 5.4 Pearl millet seed from Tabi.

April 2, 2010. Sitting under the shade of a veranda, whose roof is packed with a 40 cm thick layer of millet stalks (for animal fodder), the protection from the scorching sun is highly appreciated while talking with a group of 12 farmers in Tabi village, Dogon country. Asked to tell a story about their local millet seed, one farmer starts: 'In the past, a man from our village once ended up without any millet and had eaten his seed saved for the next season. He felt so ashamed that he did not dare to tell the others in the village. He went to Boni market and bought millet from San (a few hundred kilometres south). Most of us buy pearl millet for food at the market from time to time, but never for seed. He decided to keep quiet and used the millet he bought as seed. At first the plants grew well, but by the time all villagers were harvesting their crop his pearl millet had not yet even flowered and it wilted when the rains stopped. Not only had he wasted the millet, but he also ended up harvesting nothing.'

Dogon farmers integrate seed and grain. Seed is selected from the grain, usually before harvest, and may or may not be stored separately. Seed panicles are cut with a piece of the stem attached (about 10 cm), and can always be distinguished from panicles destined for grain, which are cut at the very base of the panicle. Most farmers store pearl millet on the panicle, and only thresh it when offering it for sale at the market. Farmers are so determined to preserve their own seed that, after 2 years of near crop failures, 80% said that they still had sufficient pearl millet seed for sowing (CRS, 2006).

Many farmers in Tabi village produce surplus pearl millet and store grain for several years, also for use as seed, as crops fail one out of three years in this region. Also, even in years where the crop does not fail but drought hits during the grain filling phase, the grain harvested will not be well filled and thus will not have good vigour at germination.

Infrastructure. Dogon farmers have continued to build storage structures like the ones made by their ancestors, some of which are still in use. Seed is often stored with the food grain in a house-like, rectangular adobe structure (roughly 6 m × 2 m × 2 m). In Tabi, it is locked behind a small iron door, to guard the precious seed stock. The structure is raised half a metre above ground, resting on piled rocks to keep out termites.

Linkages. Following 2 years of crop failure, those farmers who did not have enough seed said they relied on local grain markets as a source for local variety seed.

In 2006 and 2007 ICRISAT with local partners conducted farmer field schools (FFS) for controlling striga weed. Ever since, farmers in Tabi have continued thinning (removing the weak plants from pockets at an early crop stage), as they said this practice helps to reduce the weedy *chibra* in their crop.

Quality control. The neighbouring villages Tabi, Tega and Toupere only grow the pearl millet of Tabi. Cross-pollination by wind and exchanges between farmers have helped to maintaining their local landrace. To avoid contamination, farmers were not interested in testing new varieties of pearl millet as part of the farmer field school programme. The social cohesion, norms and rules (not to mention the demanding growing conditions) are such that farmers only source seed from within their own community.

Panicles are usually selected in field before harvest, but this may also be done after harvest by the older people. They choose panicles that are firm when pressed between thumb and index finger, and have good seed set and well-filled grains, especially covering the tip. These characteristics also prevent the grains from being damaged when rain is still falling during harvesting time.

When selecting seed, farmers identify panicles that have well-covered tips, like the one shown on the left.

5.6.3 Cash flow

In the early 1980s a project operating near Gao visited Tabi village various times with a truck to buy pearl millet for distribution in their target zone. Before that Tabi farmers were already selling pearl millet at the markets of Boni and Simby, although exchanges and sales within the village were more important. As farmers only sell grain (selling seed being culturally inappropriate as one is supposed to help a farmer in need) and leave it up to their clients whether they want to use it for food or for seed, there is no special price for seed. Prices are slightly higher in the market, to cover farmers' travel costs.

But, as the demand for their pearl millet increases, at planting time farmers sell it 30% to 50% above the grain price to outside buyers, such as to the NGO Afrique Verte, which bought 10 tonnes of millet at 250 FCFA ($0.55) per kg in 2003.

5.6.4 Marketing

Farmers confirmed that pearl millet from their villages is very much sought after by other farmers in the surrounding areas. They mentioned that people from as far away as Timbuktu, Kidal, Burkina Faso and Ségou come to their village to buy pearl millet seed. When approached for selling pearl millet, the village committee meets. Rather than coordinating the sales, they pass on the information to all village members and each decides, based on their available stock, how much they will put up for sale.

The Catholic Relief Services (CRS) also encouraged pearl millet surplus producers from Tabi and other villages to market their own stored grain to improve seed security for disadvantaged farmers in Douentza through seed fairs and vouchers.

In the Boni market in the Haïre commune even traders dealing with large volumes of grain, often imported from southern Mali, stored grain from local pearl millet varieties destined for sale during the sowing period, when farmers may use it as seed. These traders said they prefer buying grain from the group of villages around Tabi, Tega and Toupere, as their variety is very much appreciated in the area (CRS, 2006).

Dogon farmers would rather sell their belongings to survive than eat their millet seed.

When mentioning our visit to Dogon country, Faso Kaba (Section 5.2) took immediate action to establish links to source this early maturing pearl millet. It would make sense (from a technical, economic and marketing point of view) for traders to sell seed packages under the name of the village of origin, e.g. as torrinion millet of Tabi, but this option may be influenced by the rigidity with which the new seed law will be implemented.

5.7 Challenges and Strengths of the Seed Enterprises

The seed sector is changing fast in Mali. Most of the enterprises in this chapter started in the last 5 years, with both the government and donors trying to create an enabling environment.

Faso Kaba is successful because of the know-how and the ability of the owner, who has formed partnerships with research, extension, NGOs, cooperatives, individual farmers and the media. The company built a solid network of seed producers, contact farmers and farmer-customers, cleverly combining mobile phone and local radio to assess demand (for other examples of entrepreneurs who combine social networking with mobile phones, see Mehta *et al.*, 2009). Over the years Faso Kaba developed trading expertise and, as suggested by a donor, also started producing seed. Although Faso Kaba has grown and is gradually establishing its production and distribution networks, it believes that more appropriate bank loans are needed to advance its business.

According to the supervisor of Niégué Farm, demand for rice seed is growing rapidly. Meeting this demand requires more land and machinery and more off-season production. But there is no way to expand the farm without buying or renting land far away. During the rainy seasons Niégué Farm floods and the farm can do nothing about it, because the drainage system needs repairs and is managed by the Office du Niger, not by the farmers. The Niégué Farm is successful because the owners are agricultural specialists who work for themselves. They are serious and dream of progress, of becoming leading seed producers in Mali.

The Nipagnon Cooperative received financial, technical and marketing support from SG2000. Although most of its members are illiterate, the cooperative is able to apply for loans from the National Agricultural Development Bank and Kafo Jiginew, a rural credit organization. The Nipagnon Cooperative also exports seed to neighbouring countries. By attending the annual seed fairs organized by the Ministry of Agriculture, the Cooperative staff meet other seed producers and dealers and learn about what they are doing.

COPROSEM at Siby lacks business management skills. But they are expert farmers, from years of experience and from having participated in PVS. And they are close to ICRISAT, national research and extension, which helps farmers of the cooperative to have access to new varieties. They also link with seed distribution companies, such as Faso Kaba, a crucial approach to diffuse new varieties on a large scale (Siart, 2008). Equally important, one of the members, a trader by profession, is a keen innovator, who started preparing 100 g packages of new sorghum varieties to sell on the market.

The cases in this chapter suggest that one has to move away from seed production models and move towards seed production-marketing-enterprise models. Government and donor programmes that build self-reliance, along with practical organizational and management skills can equip producer organizations to start and sustain trade relations with companies and financial institutions (Bingen *et al.*, 2003).

In Mali's harsh and diverse environments it is crucial to get the variety right. The Dogon case, the Nerica case and the Niégué Farm case all show this. Various villages in the arid Dogon country, such as Tabi, have developed a strong reputation for producing good quality, early maturing and drought tolerant pearl millet. Modern varieties have never outperformed the local landraces under such harsh and variable conditions. Customers have come as far as 600 km to buy pearl millet from Tabi, while CRS has used it in seed fairs during times of food insecurity. Recently, Faso Kaba also became interested in marketing their seed. It may require Faso Kaba to expand its range from certified to truthfully labelled seed, if the seed law allows such latitude.

The new seed law, which is nearly approved, will forbid sales of uncertified seed. Seed legislation based on European models often fails to respond to African realities and needs. Pearl millet is especially difficult to produce within the norms and can often only maintain its properties when produced in the place of origin (Christinck, 2002). The registered pearl millet variety Toronion is actually a landrace (that locals call *torri yo* or *torri nyo*) originating from a village in the Mopti region and which was purified by IER. Whether the pearl millet produced in its area of origin is still the same as the registered variety is questionable. Also, field inspections for certification in the remote Dogon country would make little economic sense. The extent to which expert pearl millet farmers like those from Tabi could play a role in the national economy will depend a lot on how pragmatically the new seed legislation will be interpreted.

Governments may give entrepreneurs room to grow by supporting certification, organizing national seed fairs and supporting the financial sector, and yet policies and overly restrictive legislation may keep some community initiatives from reaching their full potential.

References

Bingen, J., Serrano, A. and Howard, J. (2003) Linking farmers to markets: different approaches to human capital development. *Food Policy* 28, 405–419.

Catholic Relief Services and Partners (CRS) (2006) *Evaluation of Seed System Security Assessment (SSSA) Douentza, Northern Mali. March 2006. Final Report.* Catholic Relief Services (CRS), Bamako, Mali.

Christinck, A. (2002) *'This Seed is Like Ourselves'. A Case Study from Rajasthan, India, on the Social Aspects of Biodiversity and Farmers' Management of Pearl Millet Seed.* Kommunikation und Beratung 47. Margraf Publishers, Weikersheim, Germany. 189 pp.

Dembélé, N.N., Staatz, J.M. and Weber, M.T. (2003) *Impact of the Malian Cereals Market Reform Program on Farmers.* International Development Policy Syntheses 68. Department of Agricultural Economics, East Lansing, Michigan State University.

Diakité, L., Sidibé, A., Smale, M. and Grum, M. (2008) *Seed Value Chains for Sorghum and Millet in Mali. A State-based System in Transition.* IFPRI Discussion Paper 00749. IFPRI, Washington. 34 pp.

Mehta, K., Semali, L., Maretzki, A., Findeis, J., Kuldau, J. and Willyerd, K. (2009) Cell phones and social networks: defining new opportunities and discovering champions for entrepreneurial ventures in developing communities. Unpublished MS, Pennsylvania State University.

Siart, S. (2008) *Strengthening Local Seed Systems. Options for Enhancing Diffusion of Varietal Diversity of Sorghum in Southern Mali.* Kommunikation und Beratung 85. Margraf Publishers, Weikersheim, Germany. 185 pp.

Smale, M., Diakité, L., Dembélé, B., Traoré, I.S., Guindo, O. and Konta, B. (2008) *Trading Millet and Sorghum Genetic Resources. Women Vendors in the Village Fairs of San and Douentza, Mali.* IFPRI Discussion Paper 00746. IFPRI, Washington. 40 pp.

Warms, R.L. (1994) Changes in attitudes toward business and economy in West Africa. *Small Enterprise Development* 5(3), 46–52.

Weltzien, E., Christinck, A., Touré, A., Rattunde, F., Diarra, M., Sangaré, A. and Coulibaly, M. (2006) Enhancing farmers' access to sorghum varieties through scaling-up participatory plant breeding in Mali, West Africa. In: Almekinders, C. and Hardon, J. (eds) *Bringing Farmers Back into Breeding: Experiences with Participatory Plant Breeding and Challenges for Institutionalization.* Agromisa Foundation, Wageningen, the Netherlands, pp. 65–75.

World Bank (2008) Mali rural finance report. Unpublished report. World Bank, Washington. 59 pp.

6 Guinea: Networks that Work

FLORENT OKRY, DANIEL N. DALOHOUN,
SÉKOU DIAWARA, M. BILLO BARRY
AND PAUL VAN MELE

6.1 Introduction

6.1.1 Agriculture

Bordered on the north by Senegal and Mali, on the south by Sierra Leone and Liberia, on the east by Côte d'Ivoire and Mali and on the west by the Atlantic Ocean and Guinea Bissau, Guinea has one of the most favourable climates for agriculture in West Africa with 6.7 million hectares of cultivable land, of which only 24% are farmed. Although mining (of mainly bauxite, aluminium, gold and diamonds) is the country's major source of foreign exchange, for the 9.7 million citizens agriculture is the major occupation. It contributes to 18.7% of the gross domestic product (2004). Rice, maize, sorghum, millet, cassava and fonio are common food crops, with coffee, rubber, palm and cocoa as the main cash crops. Shifting cultivation is common.

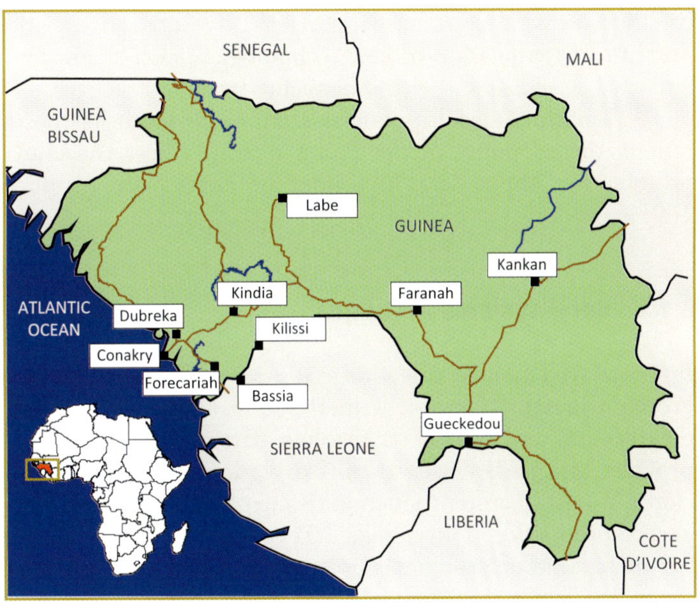

Guinea has four natural regions, defined largely by rainfall. Lower Guinea receives 2800–4000 mm spread over 6 months. Upper Guinea is drier with 1300 mm of rain over 5 months. Middle Guinea (Fouta Djallon) is slightly wetter with 1500 mm per year, whereas Forest Guinea receives 2000–3000 mm of rain during 9 months.

Rice is the most important food crop and its production is the most organized. In 2000, rice production covered 42% of the total farmed land (about 700,000 hectares) for a total production of 700,000 tonnes of paddy (Barry, 2006). Rice is grown in all four natural regions of Guinea. About 65% of rice land is devoted to upland rice, followed by mangrove rice (16%). Lowland and flooded rice are of equal importance. Per capita rice consumption is estimated at 69 kg (WARDA, 2007). In 2003, the local rice sector generated about 340 billion Guinean francs (GNF) ($67 million), 5% of the gross domestic product.

Before the 1950s Guinea was the third largest rice producer in Africa, after Egypt and Madagascar (Portères, 1966). Guinea was self-sufficient in rice and exported a surplus to other countries in West Africa. In recent years, population growth (3.1% per year) has threatened Guinea's food security. Despite a doubling in production over the past decade, to 1.47 million tonnes in 2009, rice is now imported to meet the rising demand. Imports were estimated at 44% of the national rice demand in 1995, falling to 25% in 2000 and rising again to 40% in 2002 (MAEF, 2007a).

Guinea used to export rice, but now has to import it to meet the rising demand.

To increase food security, the Guinean government plans to introduce rice cultivation on 25,000 more hectares in favourable areas of Lower and Upper Guinea (MAEF, 2007b). It wants to boost the production to 2.5 million tonnes by 2015. To achieve this, the government collaborates with international partners, invests in roads, bridges and dykes and supports the dissemination of new technologies, such as improved rice varieties (Nerica and the CK series) and yield-enhancing farming practices.

At a national level, potato is a minor crop but has become increasingly important in the economy of the Fouta region, where the climate is favourable to its production. About 16,000 tonnes of potato are produced every year (MAEF, 2007a).

6.1.2 Seed systems in Guinea

Informal seed system. The informal seed system supplies the bulk of seed to farmers. From the previous harvest, farmers and local seed dealers save seed for the next cropping season, and pass it on through barter, gift or sale. The informal seed system provides inexpensive seed thanks to its low production cost. Seed is produced and stored as part of crop production (Richards, 1986). However, a few farmers specialize in seed production (Okry *et al.*, 2011). The informal seed system is more effective at supplying seed of orphan crops (such as fonio, yam and potato) and self-pollinated crops. In Guinea the informal seed system supplies more than 90% of farmers' seed (SNPRV, 2001).

Formal seed system. From production to sale, formal seed is broken into discrete activities, done by different stakeholders rather than a single farmer, and it is fully regulated by the government. The Institut de Recherche Agronomique de Guinée (IRAG) conducts breeding for all crops and produces breeder and foundation seed, supported by selected farmers and farmers' associations. The national extension system (Agence Nationale de Promotion et de Conseil Agricole – ANPROCA, ex-SNPRV) then distributes foundation seed to farmer seed producer groups, who multiply it into quality seed, under its supervision (only when projects support the activities). Seed producers, some of whom are organized in associations or cooperatives, usually sell the seed themselves. There are only a few retailers in Guinea. The formal seed system focuses exclusively on improved varieties and commercial crops, such as cotton and cocoa.

As the seed production units are located near cities, farmers in remote areas are discouraged from accessing quality seed. Moreover, many are reluctant to pay more than the grain price to buy seed if they are not sure the source can be trusted or if they are unaware of the added benefits of the quality seed.

6.1.3 Evolution of the formal rice seed sector in Guinea

The first support for the formal seed sector was in 1986, as part of a broader food security programme funded by the World Bank. Four well-equipped seed centres were built in Kilissi, Koba, Guéckédou and Bordo to produce, process, store and package 'acceptable seed' under the supervision of the national research system (acceptable seed has followed the standard procedures for seed production, but has not been certified). Two early maturing and improved varieties, CK 5 and CK 7, were identified to increase national rice production.

The newly created seed centres multiplied foundation seed, processed and packaged resulting seed into 5 kg bags and developed, in collaboration with SNPRV, technical notes related to the characteristics and use of each variety. SNPRV disseminated the bags free of cost to selected farmers, who were trained and expected to diffuse seed and technical information within their community. The programme ended in 1992. Farmers had not accepted the two improved varieties: they were too early maturing (requiring intensive bird-scaring and delicate postharvest processes) and too sensitive to weeds.

The seed centres closed in 1997 because of limited impact and lack of funds. They remained government property until the liberalization of the seed sector in 2004, when they were handed over to private entrepreneurs and farmers' associations (Okry *et al.*, 2011).

In 1997 IRAG and the extension service launched a pilot programme with AfricaRice (ex-WARDA) to accelerate the diffusion of Nerica (varieties created by crossing two rice species: *Oryza glaberrima* × *O. sativa*). At first, four of these interspecifics were introduced in Guinea: Nerica 3, Nerica 4, Nerica 6 and IAC 164. Later projects, such as the African Rice Initiative, continued to disseminate Nerica varieties.

Meanwhile, from 1997 to 2003 Sasakawa Global 2000 (SG2000) extended its activities to Guinea, organizing small-scale seed producers. Apart from rice, SG2000 promoted cowpea, maize, soybean and the cover crop mucuna (velvet bean), to enhance soil fertility (SG2000, 2005).

More often, seed components have been part of agricultural development projects that promoted improved varieties and created farmer seed enterprises, which collapsed once the projects ended.

6.1.4 Seed legislation in Guinea

Seed law. Guinea, with the FAO, wrote a law on seed and agro-inputs, applicable to all agricultural crops. It sets the norms and aims to stimulate production, marketing and the use of quality seed. It also stipulates who can certify seed and how it should be done. The law was recently approved by parliament and will soon take effect.

Seed certification. The Direction Nationale de l'Agriculture (DNA) should implement national seed policy and seed quality control, but the quality control laboratories are under-equipped and non-functional, the national seed committee is still not formed and the seed inventory is not yet completed. Moreover, there is not enough staff trained in seed certification. In reality, there is little proper seed quality control. Most seed producers simply produce at standards acceptable to their clients.

Below we present five seed enterprises in Guinea. They all include rice, the major staple. The first case deals with a traditional rice seed producer and dealer who developed a small seed distribution network in two countries. She produces seed of local rice varieties without agrochemicals. Besides seed, she also sells paddy and palm oil, which at times clients exchange with her for seed. The second case is another one-person seed enterprise that started in the 1980s with project support, but which is still in business 5 years after the subsidies ended. The third case is an agro-dealer who also sells seed of rice, maize, cowpea and vegetables. The last two cases describe enterprises that mainly produce seed potato, followed by rice and maize.

6.2 Mama Adama Yansané

6.2.1 History

Mama Adama Yansané, a 60-year-old widow, has been running her rice seed business for over three decades. In the 1970s she started trading rice seed in Bassia, Sierra Leone (Bramaia chiefdom), where she got married. A few years after her husband died, she returned home to Bokariya-Tassen in Guinea in the 1990s.

After returning Mama Adama continued running her one-woman seed enterprise, and kept taking seed orders from Bassia and the surrounding area. She married the village imam in her home village. Unlike neighbouring villages, selling seed is forbidden in Bokariya. The imam banned it, citing religious reasons (Box 6.1). Mama Adama's seed business is informal (non-registered) and small, even though it operates in two countries. Her seed business steadily grew until the Sierra Leone civil war

Box 6.1 The unspoken profession.

When we first asked about seed dealers, people said that there were none in their village, because the imam had forbidden it. As we did more interviews, we realized that many farmers buy seed from Mama Adama. One interviewee even requested anonymity if we further discussed Mama Adama's seed business with others.

When we met Mama Adama she readily admitted that she sold rice seed, but added that her clients were mainly from outside the village. She said in the village she sells rice to people, not seed. When we asked, 'What is the difference between the seed sold to outsiders and the rice sold to Bokariya farmers?' she admitted that there was no difference. Mama Adama was creatively bending her husband's rule to create a seed monopoly.

started in 1991. Her clients left Bassia to settle elsewhere and farmers from villages across the border in Guinea also began growing less rice, for fear of invasion.

Mama Adama has always specialized in seed of local rice varieties. Before the war in Sierra Leone ended in 2001, she grew and sold *dixi* (*O. glaberrima*) and *samba* (*O. sativa*). However, she abandoned these varieties after their demand dropped. *Dixi* is a 'heavy variety', meaning that people who eat it feel full for a long time. *Dixi* was a favourite during the war, but it is sensitive to drought during flowering and its popularity declined. *Samba*, on the other hand, is a 'light variety' and is mainly eaten by elderly people, given to children at boarding school or sold to people in the cities. After Guinean farmers who had fled the border area returned home, they brought with them a new variety called *saidou gbéli* (*O. sativa*) named after the person who introduced it, although no one is sure where *saidou gbéli* came from. It is high yielding, drought-resistant, adapted to the local environment and suitable for both uplands and lowlands. Besides, it can be sown from late May to early September, unlike all other varieties, which have narrower planting windows. Mama Adama now sells only *saidou gbéli*, as it soon became the most popular variety in her area.

Mama Adama Yansané stores her rice seed in a basket, while she keeps paddy in bags.

Every year Mama Adama produces a field of rice seed, but she does not keep records of her sales. She has always sold her entire seed production (in 2007 this was 1.5 tonnes from 1.2 hectare). Her seed business prospered after the end of the war, as demand for seed outstripped supply.

Mama Adama says that the demand grows every year as larger areas are planted to rice, while she produces less seed as she gets older. Mama Adama could enlarge her business by letting someone else grow the seed or by buying seed from other farmers, but she is afraid of compromising quality. She thinks that seed quality, and her reputation, should be preserved by all means.

6.2.2 Structure

Management. Mama Adama's seed business serves neighbouring farmers. She has no contact with external organizations working in seed development, although she took a literacy course from APEK (Association pour la Promotion Economique de Kindia), a local NGO.

Besides rice seed, Mama Adama sells groundnut seed, palm oil and rice paddy. She started trading palm oil and paddy when she started her seed business. These three enterprises are linked because palm oil and paddy are often bartered for rice seed.

Paddy and rice seed are grown, stored and handled separately. Mama Adama grows and processes seed by hand, using no pesticides or botanical products to control storage pests. Seed is stored in a large basket, while paddy is stored in polythene bags.

Land. In Bokariya-Tassen, land is managed communally. Every year a committee of elders allocates land to each household. Allocation within the household rests with the household head, so Mama Adama depends on the imam for land. Once she receives her land from the imam, she selects the best patch to grow seed. She knows that the most fertile land has more shrubs, darker soil and greener vegetation. Her husband usually gives her the land she desires.

Labour. Land preparation, weeding and harvesting are the most labour-intensive tasks in rice farming in Bokariya. Finishing them on time determines the success or failure of one's rice crop. At these crucial times labour is scarce as the demand is high and only people with special relationships with the work crew can hire labour then. Villagers who previously received a gift or a loan of seed from Mama Adama may give her priority when deciding who to work for, out of respect.

A passion for quality and a focus on a single variety have built Mama Adama's reputation.

Quality control. Mama Adama believes that quality comes from keeping control of her seed production at all times. 'The quality of my seed keeps me in the seed business. People come from far away to buy my seed. How can I assure quality if I do not produce my own seed?' She added, 'I have kept only one variety to avoid seed mixture.'

Keeping seed pure requires skill and work. To avoid mixture, Mama Adama does not harvest seed from the edges of her seed field. These areas are harvested as paddy. Since farmers broadcast their seed and rice fields are continuous with no clear borders between neighbouring fields, edges can easily have a mix of varieties. To prevent seed from mixing with other varieties and dirt and to reduce time for scaring away chickens and other rice lovers, Mama Adama dries seed under the sun on tarpaulins in her courtyard.

She tests seed dryness by cracking a few grains between her teeth. The sound of the cracking grains tells her the dryness of the seed. She believes that seed that is harvested on time, well-dried and properly stored has a good germination rate. She does not need to test that.

6.2.3 Cash flow

Apart from selling seed, Mama Adama also buys paddy from farmers just after harvest, stores it and resells it as paddy throughout the year. At the beginning of the season seed and paddy cost the same, namely 2000 GNF per kg ($0.50). Quality seed (clean, insect free, unspotted and well-dried) becomes more expensive and can reach up to 3500 GNF per kg ($0.90) towards the end of the planting season. The price of paddy also varies throughout the year, but never reaches this level.

Farmers buy seed or barter for it. Mama Adama may give or loan seed depending on the person involved. Some farmers also trade paddy for seed. The exchange rate is not fixed.

Mama Adama runs her business with her own funds. She could borrow from local moneylenders, who often charge up to 100% interest, but she believes that no credit service will give her a loan since she is too old.

Mama Adama produces seed without agrochemicals, thus reducing her costs. All farm tasks are done by hand, for which she hires labour.

6.2.4 Marketing

Most of Mama Adama's customers are rice growers, informal dealers and occasionally farmers' associations (Table 6.1).

Table 6.1. Clients of Mama Adama Yansané.

	1995	2000	2005	2009	2015 (predicted)
Individual farmers	1	1	1	1	1
Local dealers	2	2	2	2	2
Groups and cooperatives	–	–	–	3	3
Projects and NGOs	–	–	–	–	4

Ranking assessment by senior management of seed enterprise, 1 being the most important.

Her seed network expands through her kinship ties. For example, Samba Yansané, one of her nephews living in Bassia, Sierra Leone, takes seed orders before the sowing season from nearby villages: Siakhaya, Kabaya, Sabuya, Surumaya, Fadugu, Sogbaya, Yaya and Sulemania. Samba's father used to take the rice orders before he died. Samba, only 22 years old, travels to Bokariya, crossing the Great Scarcies River on the border, to get seed from Mama Adama. Samba collects transportation fees and receives a commission from farmers. In Guinea, customers come from Kaff (7 km), Sangaran (1.5 km), Konkoya (1.5 km), Boubouya (4 km), Salamou (4 km) and Kondedara (5 km). Although Mama Adama occasionally used to take seed to her relatives, now most customers come to her house to get seed.

Mama Adama only has so much flexibility. She has a fixed capacity to produce seed and produces none off-season. The seed she sells is from the year before and, as she fears enlarging the enterprise, the quantities have not really changed. Once her seed is sold she never buys paddy to sell as seed.

Mama Adama does not pack her seed before selling it. Customers come with their own containers, mostly plastic sacks, or a piece of cloth and a bowl for small quantities.

6.3 Ibrahima Sherif

6.3.1 History

Sherif, 66 years old and a driver by profession, was born in Samoukiry, Lower Guinea. In the 1980s he migrated to Foulaya, where he started farming. He was growing avocado trees, vegetables and a bit of upland rice when a delegation from the World Bank visited his village with government extension agents to monitor one of their projects in 1984. He must have made a good impression because soon after the visit the director general of the extension service allowed him to get a lease-to-own for a water pump to increase his production. The charges amounted to 500,000 GNF ($100). With the new equipment, Sherif increased his production and became a model farmer, regularly receiving and impressing official visitors.

When donors changed their policy on agricultural extension in the 1990s and group approaches came into vogue, the local NGO APEK started grouping farmers into associations and unions. Sherif was elected chair of the union of cereal producer groups of Kindia. In this role Sherif negotiated a tractor for the union, but of course he could use the tractor too, and so he increased his production and with the surplus bought fertilizers, herbicides and pesticides.

In the early 1990s Sherif produced seed of local varieties, such as *sewa*, an upland *O. sativa* variety, but in 1996 he abandoned it to produce seed of improved varieties of rice, soybean and mucuna with support from SG2000.

Over the years, he has collaborated with several projects. One day the former President, the late Lansana Conté, visited Sherif's farm and soon after he received a tractor for his own farm from the Ministry of Agriculture and a water pump from SG2000. However, in 2004 SG2000 left Guinea, ending the subsidies for Sherif's seed business. After that he bought all his own inputs, although in smaller amounts.

Sherif tried growing seed of the upland rice Nerica 4 in 2003, 2004 and 2006, but abandoned it because threshing was tedious. Ideally, Nerica 4 should be threshed the same day it is harvested. When the bundles remain in the field for a couple of days, it has to be threshed by machine. Sherif focuses on seed of improved lowland varieties, such as CK 90, CK 21 and CK 801 (Table 6.2). Occasionally, he grows Nankin. All these are improved lowland rice varieties. Nankin was introduced by the Chinese, while the CK series were bred by the national rice breeding unit at the Kilissi station. These varieties have yield potentials of 5 to 6 tonnes per hectare and are all found in the local seed trade.

Table 6.2. Rice seed produced (tonnes), Ibrahima Sherif.

	2005	2006	2007	2008	2009
CK 90	0.3	5.0	15.0	–	–
CK 21	5.1	5.0	9.0	2.0	1.0
CK 801	0.1	1.0	5.0	1.0	5.0
Dia	–	–	–	–	3.0

In 2008, Sherif decided to include *dia*, a local rice variety, for seed production. He grew 3 tonnes of *dia* in 2009 and plans to include more local varieties in 2010. Sherif began growing local rice varieties when he had to produce seed without subsidies and also to meet smallholder farmers' explicit demands.

6.3.2 Structure

Management. Like Mama Adama Yansané, Sherif runs a one-man seed enterprise, but, unlike her, he has collaborated with several seed projects that led him to produce improved varieties, besides giving him much valuable equipment.

Besides rice seed, Sherif has produced smaller amounts of seed of maize (since 1997), cowpea (since 2000), soybean (2000) and mucuna (1997–2003), as requested by SG2000. But he stopped producing mucuna seed because of the limited demand. For years, mucuna was promoted as a miracle crop to restore soil fertility, but farmers never took it up because, as they correctly say, it occupies cultivable land and it is not edible. After he stopped receiving subsidies he reduced the area grown, cut down on the use of agrochemicals and included local rice varieties in his portfolio.

Land. Sherif owns about 46 hectares, of which 44 hectares are lowlands and 2 hectares upland. Only 41% of the land is exploited. Upland fields are used for maize seed (1 hectare) and cowpea seed production (1 hectare). About 17 hectares of lowlands are devoted to lowland rice seed production. Seed fields of different varieties are far away from paddy fields and separated from one another by dykes to reduce the chances of seed mixture by flood.

Sherif had to include local rice varieties and gradually built his network after projects came to an end.

Equipment. Since 2003 Sherif has owned a tractor, which is still running. He also has a pair of cows for ploughing and two water pumps for irrigating occasional off-season seed.

Labour. Sherif has limited household labour. He hires labour for land preparation, building dykes, transplanting, harvesting and threshing. The household labour takes care of other crops like fonio and vegetables.

Links. Sherif has extensive relationships with rural development organizations. For many years, research and extension have used Sherif as a model farmer.

Quality control. Sherif receives no external quality control for his seed farm. He has participated in several training sessions on seed production in Guinea and Senegal. These courses and follow-up sessions from research and extension have helped Sherif to do his own quality control.

6.3.3 Cash flow

Since SG2000 left in 2004, Sherif has run his seed business on his own funds. The equipment he obtained from SG2000 and the Ministry of Agriculture helped. However, Sherif realized that he cannot keep producing only seed of improved rice varieties, which he once sold mainly to projects and farmers' unions. To stay in business and reduce production costs he also started producing local rice varieties in 2008.

To individual rice growers, Sherif sells seed on average at 3500 GNF ($0.90) per kg, but the price can reach 5000 GNF ($1.30). Unlike many seed dealers in Guinea, Sherif no longer accepts loans, gifts or barters. 'Some farmers reimbursed their loan with mixed seed and some even did not reimburse me at all,' he said.

Customers referred by the Chamber of Agriculture (Chambre d'Agriculture) bought seed at 2500 GNF ($0.70) per kg. Sherif said the Chamber of Agriculture helped him acquire inputs and equipment, so he gives them a special price because he wants to keep good relations with them.

Sherif said he sells rice seed for just 2000 GNF ($0.50) per kg to farmer groups and unions as an expression of solidarity with them, even though he is no longer the chair of the farmers' association, and to show commitment to the groups' efforts to raise revenues of other farmers. Sherif also uses the farmers' union to develop his seed distribution network. He believes that by offering a discount to groups he will encourage individuals to try improved varieties.

6.3.4 Marketing

Sherif's clients have changed a lot over time, as he gradually started to build his network after projects came to an end (Table 6.3). Local dealers are likely to play a stronger role in the future.

Table 6.3. Clients of Ibrahima Sherif.

	1995	2000	2005	2009	2015 (predicted)
State farms	2	2	2	1	4
Individual farmers	–	3	3	2	1
Groups and cooperatives	–	4	4	3	2
Research institute	–	6	6	4	6
Local dealers	–	5	4	5	3
Projects and NGOs	1	1	1	6	5

Ranking assessment by senior management of seed enterprise, 1 being the most important.

Up to 2004, customers were mainly referred by the Chamber of Agriculture (50%), followed by individual rice growers (30%) and farmers' unions (20%). The Chamber of Agriculture did not put official orders for its own use, but directed seed orders from projects, state farms and NGOs to Sherif as a faithful, recognized seed producer.

Seed is sold at Sherif's store. The discount Sherif gives to associations is part of his marketing strategy, as individual farmers get to know him through this. If off-season seed production is abundant, Sherif advertises at the local radio station before farmers start sowing. Advertisement usually takes 4 weeks and starts about 2 weeks before the sowing period. The radio spot explains the importance of improved varieties, and says that Sherif produces and sells the seed, and that he can be reached at his village, or through farmer's associations, researchers and extension agents. The advertisements helped Sherif build his popularity and reputation in the area.

6.4 Comptoir Agricole

6.4.1 History

In the early 1990s the Belgian NGO ACT (now called TRIAS) invested heavily in training farmers and building roads to improve farmers' access to markets. It noticed that farmers of Bangouya village, near Kindia, could make better use of their lowlands if they had vegetable seed, which ACT helped them acquire.

In 1994, before ACT stopped its intervention in Guinea, it decided to organize Bangouya's farmers in a cooperative called CCIAK (Coopérative de Commercialisation des Intrants Agricoles de Kindia) to obtain vegetable seed and agro-input supplies. It was led by former ACT staff. Incofin, a Belgian social investment company focusing on micro-finance, provided financial support for CCIAK and helped them import vegetable seed from Belgium. CCIAK rapidly enlarged its activities and started a store in Kindia.

Also in 1994, Comptoir Agricole was created and registered as a private company to distribute agricultural inputs (seed, agrochemicals and farm equipment) and food (such as paddy). CCIAK and Comptoir Agricole thus shared the seed market.

CCIAK cooperated with Incofin until 2000, when the latter stopped working in Guinea. However, until 2003 Incofin linked CCIAK to Belgian and Dutch seed companies, allowing CCIAK to import vegetable seed for sale.

In Guinea's financial crisis of 2003 the Guinean franc lost value and CCIAK nearly went bankrupt. It stopped importing vegetable seed and adjusted by engaging in local seed supply. Due to the harsh financial crisis, CCIAK could not survive alone and merged with Comptoir Agricole in 2004, retaining Comptoir Agricole as its official name.

In addition to farm inputs, Comptoir Agricole sells up to 100 tonnes of rice seed per year, although it does not produce any of the seed itself (Table 6.4). It also sells seed of maize and groundnut and resumed its former trade in vegetable seeds (tomato, pepper, cucumber and onion). The amount of seed sold per year largely depends on the money available to import vegetable seed and the amount of good local seed that its scouts can source.

In 2008 and 2009 Comptoir was hit by the global financial crisis. To reduce overhead costs it got rid of salaries by handing over its seven shops to its employees. Currently, it supplies the shops with inputs. After sale, Comptoir deducts the capital and leaves the profits to the employees, who now earn money only if they sell.

Table 6.4. Seed sales (tonnes), Comptoir Agricole.

	2007	2008	2009
Rice	98	78	67
Maize	5	7	6
Groundnut	40	3	2
Vegetables	0.3	0.7	0.5

Individual rice growers from Madina Oula, near the southern border with Sierra Leone, supply Comptoir Agricole with seed of both local and improved varieties. Some seed comes from government employees who farm their own land part-time. Occasionally Comptoir Agricole sells seed of improved varieties produced by the national rice breeding unit at the Kilissi Research Centre (Table 6.5).

Table 6.5. Rice seed supplied (tonnes) to Comptoir Agricole.

Supplier	Variety supplied	Quantity (tonnes)*
Formal seed producer from Labe	Nerica 4	6
Kilissi Research Centre	CK21, CK90	50
Rice growers	Nankin, *kaolaka, saidou gbéli, saidou firê*	80
Sunday farmers (loan reimbursement)	Diverse	1.5

*Data from 2007; includes both seed sold and emergency seed distributed under the FAO project Office of Special Relief Operations (OSRO).

The rice varieties sold include *saidou firê* (local variety), *saidou gbéli* (local variety), *kaolaka* (local variety), Nankin (improved variety), Nerica 4 (interspecific), CK 21 and CK 90 (improved varieties). *Saidou firê* and *saidou gbéli* are the most popular upland varieties grown in Kindia region, where Comptoir Agricole has its headquarters.

6.4.2 Structure

Comptoir Agricole was created by a group of former civil servants and registered as a private company. Comptoir Agricole has seven shops located in Lower Guinea (Kindia, Dubreka and Forecariah) and Middle Guinea (Dalaba and Labé). Comptoir Agricole covers the entire country when contracted to distribute emergency seed by projects (such as OSRO) or by humanitarian and international organizations, such as FAO and the World Food Programme (WFP).

Links. Comptoir Agricole is a member of professional associations, including the national Association des Producteurs Importateurs Distributeurs d'Intrants Agricoles (APIDIA) and the international African Seed Trade Association (AFSTA).

At the local level, however, Comptoir Agricole is poorly linked with formal seed producers except for one seed producer trained by SG2000, who sold them 6 tonnes of Nerica in 2007. According to its director, Mr Hamidou Diallo, 'seed projects and local NGOs have helped a lot in training formal seed producers, but, after the training, the formal seed producers served exclusively the projects and NGOs involved. These seed growers were not allowed to diversify their clients. The projects and NGOs are trainers, suppliers of foundation seed and buyers of the produced seed.'

6.4.3 Cash flow

Comptoir Agricole sells agrochemicals, paddy and seed of various crops. It is a seed retailer, but does not produce seed. Comptoir Agricole collects seed at harvest, stores it and sells it at planting season. It does not process seed. Young people who work part-time for Comptoir visit the seed-producing villages on motorbikes and advise the company when the seed is ready. Over the years these young people have built up their social networks. They know exactly which farmers produce good quality seed. Comptoir then goes to the villages, collects the seed and pays cash at 1000 GNF ($0.25) per kg. Comptoir only sells seed on credit to civil servants farming their own land part-time. Interest rates vary according to the person involved and other chemicals bought.

6.4.4 Marketing

Farmers make up 65% of Comptoir's customers. Others are projects and Sunday farmers, who are part-time farmers. Comptoir Agricole reaches its customers mainly by participating in fairs and distributing booklets presenting its products. Since its creation in 1994, it has advertised only once on the local radio station, in 1994. Farmers are their main customers, which is unlikely to change in the future. As various efforts currently aim at strengthening groups and cooperatives (in terms of organization, training and land management) these will likely become more important clients (Table 6.6).

Table 6.6. Clients of Comptoir Agricole.

	1995	2000	2005	2009	2015 (predicted)
Individual farmers	1	1	1	1	1
Sunday farmers	2	2	2	2	3
Projects and NGOs	–	–	3	3	4
Groups and cooperatives	–	–	4	4	2
Local dealers	–	–	–	–	–

Ranking assessment by senior management of seed enterprise, 1 being the most important.

6.5 Cereal and Potato Seed Producers' Union

6.5.1 History

After many years of experience in working with farmers, as extension officers, four agricultural engineers and two agricultural technicians decided to set up an organization to produce seeds in the Fouta region, in middle Guinea. Although they still receive their basic salaries, the government barely provides an operational budget for extension. Tired of being idle, they looked for an opportunity to make best use of their time and expertise. One of the agronomists noticed that investing in agriculture is profitable but that seed production is even more so. Knowing that good seed improves yields and crop quality, the agronomist discussed the idea with some colleagues and farmers and they decided to join the existing farmers' associations as a seed enterprise.

In 2006, 19 farmers (including the six extension agents) and ten cooperatives growing cereals, ware potatoes and vegetables founded a union (Union des Coopératives pour la Production des Semences de Céréales et Tubercules). It was officially registered at Labé, the regional capital of the Fouta.

The start-up capital of the Union came from membership subscriptions. But this was not enough to start the business. Then, in 2006 a CFC (Common Fund for Commodities) project provided seed potato as a loan to the Union. After its first harvest, the Union reimbursed double the amount of seed potato. Profits from the first activities were kept and used to expand production.

The Union has rapidly increased its seed production (Table 6.7). Net production is just the part of the harvest that is selected for seed. The rest

Table 6.7. Seed produced, Cereal and Potato Seed Producers' Union.

	2007	2008	2009
Cultivated area (hectares)			
Potato	5	14	22
Maize	3	6	10
Rice	2	8	12
Net production – sold as seed (tonnes)			
Potato	17.5	55	96
Maize	4.5	12	18
Rice	4	16	36

is eaten or sold as food. Production increased because: fertile land is available for expansion; the Union hosts many cooperatives (giving them a large customer base); the Union has a good marketing strategy; and the Union's managers are extension agents who easily convince farmers about the importance of using quality seed. Apart from their agricultural skills, extension agents can often bank on the social networks built up during their lifetime, both with the farming community and with research and source seed suppliers.

6.5.2 Structure

Management and staff. The Union is composed of ten cooperatives, each with 30 members, of which 80% are women. The cooperatives were set up by women to create income generating activities. They invited some men to join.

The leaders of the Union are male agronomists who are still paid as government extension agents. Alpha Oumar Balde, who is the regional director of the extension service for the Fouta region, is the president of the Union. The other three agronomists are each responsible for one of the three crops (potato, maize and rice). Two technicians help members and ensure strict adherence to the technical procedure. None of the management staff is paid by the Union. They think that if the activity develops they may request a salary in the future, but so far they do not charge the Union, since they still draw a government salary and they do not want to burden the Union but they want to allow it to grow. Though the Union has the required technical capacities, most women are illiterate and need skill strengthening.

Production. The Union operates on part of a vast lowland of about 7000 hectares, at an altitude of 1002–1115 metres. The main crop is dry season potato. After the potato harvest, rice is planted on the lower, wetter part of the area, while maize is sown on the upper part.

Seed potato is imported from France, the Netherlands and Belgium by Sica, a private company based in Labé. Five varieties dominate: Nicola, Spunta, Anova, Kaon and Désirée. At harvest, the Union members select the smallest or medium-size tubers as seed. They renew their foundation seed for potato, maize and rice every 3 years. For maize they isolate the farm on land where they used to produce potato so no other field of maize is near it.

Activities are organized by gender. The men build the fences to protect the seed crops from goats and cattle, while the women do everything else. Traditional wooden fencing costs about 5 million GNF ($1000) per hectare. The second year, maintenance expenses are about 10% of the construction costs. The third year, there is a need to replace the fence. Using modern materials (mostly wire) the fence costs 14 million GNF ($2750), but lasts for 20 years.

Each cooperative chooses one crop, often the one they were growing before for food. No group can produce seeds for two different crops. This is to balance the supply and demand of seed. If everybody produced seed of the same crop, some crops would be under-served.

Specializing in one crop is a good arrangement for members. If seed potato is more profitable, growing it demands more work. Specialization allows each actor to master the crop and become a real professional.

It is easy for the Union leaders to identify each member's needs for new skills, so that, when training opportunities occur or are created, it is easier to designate participants.

Other activities. Each cooperative of the Union produces seed on collective land assigned for seed production, and the Union sells the seed. On individual plots each member of the cooperative has other activities including raising crops to feed their families, animal husbandry and food processing for people and animals. Cattle raising is the most common activity for Foullah (Fulani). Animals eat rice straw and maize stalks and fertilize the soil with their dung.

Women from the cooperative grow seed potato, while former extension agents joined as technical advisers and linked them to source seed suppliers and the wider farming community.

Equipment. The Union has limited equipment and infrastructure. For seed storage, it hires space from big traders of the region. It owns two motor-pumps and a few animal-drawn implements. In 2009, the Union started using the tractor ploughing services of a Frenchman to expand production.

Links and partnership. The Union has strong ties with extension and research thanks to its management. It has also developed partnerships with projects, NGOs and government bodies whom it supplies with seeds. Other farmers not in the Union are important as potential clients. The Union is trying to strengthen relationships with them and create new ones.

Quality control. The Union works under the strict control of agricultural engineers and technicians, but without certification; the seeds can be sold only in Guinea, but the Union sells all the seed it produces.

6.5.3 Cash flow

Cooperatives or individual members of the Union access some credit from small, local private credit agencies (*établissements*) at an exorbitant interest rate (4.75% per month). Fortunately the Union members need little credit, mainly for foundation seed, which they reimburse quickly. The fertile soils need little fertilizer.

Cereal seed production is rewarding, but profit from seed potato is higher. First category potato seed has a vegetative cycle of 75 days and second category one of 90 days. Sica supplies the first category at 10,000 GNF ($2.0) per kg and the second category at 6500 GNF ($1.3). The CFC project also imports potato seed and supplies it at 8000 GNF ($1.6). Ordinary farm-saved seed costs 7000 GNF ($1.4). Ware potato costs on average 3500 GNF ($0.7) per kg.

Reinvesting profits into the enterprise has been a key to success of various other enterprises presented in this book, ranging from the farmer seed producer groups in northern Cameroon who applied revolving funds (Section 3.1.4) to NASECO, one of Uganda's leading companies (Section 10.2.3). But operational budgets based on membership fees and profits are often not enough. Governments and donors have a role to play in strengthening the financial sector to support rural entrepreneurs.

6.5.4 Marketing

The Union assesses the seed demands of its members, based on the areas they intend to devote to each crop. The management of the Union approaches some NGOs, projects and other government bodies that usually buy seed from them and assesses their interest in seed for the upcoming season.

The first clients of the Union are its members, who buy up to 40% of the seed produced. The rest is sold to organizations and projects, but mostly to individual farmers. The Union gives technical assistance to the members through rigorous follow-up programmes organized by its agricultural technicians, to make sure that the members produce the seed well. This helps producers to get expected results.

Each individual member brings at least one new customer every year, a personalised marketing strategy that banks on farmers' social networks. The Union is also getting ready to use community radios to advertise.

There were demands for potato and rice seed from Senegal and efforts were made to meet them. Unfortunately, the attempt was unsuccessful for lack of certification. So far, there is no authorised body that certifies seed in Guinea.

6.6 El-Hadj Tafsir Sow

6.6.1 History

From importer to seed grower. El-Hadj Tafsir Sow was a prosperous importer of sugar, wheat flour and other foodstuffs from France. On a business trip to France in 1987, he decided to visit farms and see how wheat grows. Among other crops, he saw sugar beet, maize and potato. He was most impressed by the yields and quality of the maize and potato. He realized that good seed is important. He was amazed because he used to produce maize and potato for his family's table, but he could never imagine achieving the results he saw in France. For El-Hadj Sow this new discovery had to be shared with farmers in the Fouta (the middle Guinea region).

Sow noticed that, without quality seed and other inputs, farmers get meagre results. Once back in Guinea, he approached the research and extension service to know more about seed and to understand how it is produced. The long and demanding process requires more care and equipment, specific technical and managerial skills and knowledge about production, harvest and postharvest.

El-Hadj Sow worked with extension agents and helped researchers set up experiments on his farm to study the behaviour of many crop varieties. They also did livestock experiments. These trials became a sort of field school where he learned about seed production and farm management.

Despite all the barriers to a newcomer in the seed industry, El-Hadj Tafsir Sow was determined to take the lead in producing and distributing seed. But, when he shared his new vision with people, they told him that he could not succeed with seed production in the Fouta. But he refused to go down without a fight. So he requested Nerica rice seed from the extension service and received 29.5 kg. Today, El-Hadj Sow is happy and proud to have transformed this into more than 50 tonnes of rice seed.

Portfolio. El-Hadj Sow produces seed for the main crops of his region, namely potato, rice, maize and cowpea (Table 6.8). Though he started in 2000 with Nerica rice seed production, he could only provide figures for the last 3 years. His cultivated land remained the same in 2007 and 2008, but increased by 1 hectare in 2009. The areas devoted to seed potato, cowpea and rice seed fluctuated while maize increased. El-Hadj Sow explained that he has more land and could farm more of it but his tractor broke down and the spare parts are not available in Guinea to fix it.

Table 6.8. Seed produced, El-Hadj Sow.

	2007	2008	2009
Cultivated area (hectares)			
Potato	10	8	9
Rice	7	8	5
Maize	3	4	6
Cowpea	2	2	3
Net production – sold as seed (tonnes)			
Potato	25	17	20
Rice	18	19	10
Maize	2	4	9
Cowpea	0.6	0.5	0.8

6.6.2 Structure

Management. Except for occasional labourers, El-Hadj Sow does all the farm work with his household. One of his sons, a fourth-year agronomy student at university, gives him technical advice. He is open to new technologies and eager to transform them into innovations.

Besides crop production, El-Hadj Sow expanded to livestock. When he started his seed business, he added to his stock of cattle (as a Fulani he always has something in his animal enclosure) because he believed that crop production would be more successful when associated with animals, who can turn crop residues into manure.

El-Hadj Sow, the first potato seed grower in the Fouta region in Guinea, poses in front of his diffused-light seed potato stores, an appropriate technology.

Land and equipment. El-Hadj Sow owns 29 hectares of land. He inherited 4 hectares and bought 25. His strategy was to start by renting 3 to 5 hectares from a landlord. After farming it for 2 to 3 years he earned enough money from seed to buy the land. He now has two motor-pumps and garden hoses for irrigation, and rents many buildings to store seed and other farm products.

Links and external relations. El-Hadj Sow has close ties with researchers and extension officers, who do experiments on his farms and sell him foundation seed. When he has a technical problem he cannot solve, Sow calls upon these experts, and they always come. He has excellent relations with farmers who buy his seed. He advises his clients on how best to treat improved crop varieties. He has little or no interaction with other seed producers. Foundation seed potato importers are his suppliers.

6.6.3 Cash flow

This family farmer has more vision and ambition than capital, but no banks make agricultural loans in Guinea. This is in sharp contrast with Mali, where the government supports the Banque Nationale de Développement Agricole (BNDA) to provide farmers agricultural loans at an annual interest rate of 12% (Section 5.3.3).

El-Hadj Sow said that there are some small financial establishments but the biggest loans they give are much lower than the minimum he would need. Worst of all, they charge exorbitant interest rates, and the first payment has to be made just a month after taking the loan, long before one could harvest the crop and begin selling the seed to repay the debt.

6.6.4 Marketing

El-Hadj Sow is the first large-scale seed producer in his region. He is known for his insistence on quality seed and for strictly following the technical production system. He regularly visits his clients to encourage them and learn about their ever-changing demands.

His main customers include farmers from Labé district and state farms directed by the Chamber of Agriculture (Table 6.9).

Table 6.9. Clients of El-Hadj Sow.

	2000	2005	2009	2015 (predicted)
Individual farmers	1	1	1	1
State farms	2	2	2	2
Export in region	–	–	–	3
Projects and NGOs	–	–	–	4

Ranking assessment by senior management of seed enterprise, 1 being the most important.

Official orders for rice seed arrived at the Ministry of Agriculture from the embassies of The Gambia, Mali and Guinea Bissau in 2005 and 2006. As there is no certification, the seed could not be sold directly, but, for this special request from one government to another, quality seed was sourced from El-Hadj Sow (see above). There are still opportunities for exporting seeds to neighbouring countries, but he cannot respond to them because of lack of certification.

6.7 Challenges and Strengths of the Seed Enterprises

The formal seed sector in Guinea is still fragile. The seed laws are not enforced and regulation and certification agencies are weak. Most seed is informal, supplied by individual seed producers, farmers' associations or cooperatives. A few of them are formally registered. Seed entrepreneurs mostly supply good quality seed rather than certified seed.

Over time, the seed enterprises described above have learned about their markets and how to adapt to a changing environment with small and unpredictable seed demands, no subsidies, lack of credit and farmers who are often reluctant to pay extra for improved seed (e.g. because they do not know the variety, are unsure about the quality or believe they can save seed of a similar or even better quality). Thus production costs largely determine viability of seed enterprises. Sherif used to receive subsidized inputs, but now pays full costs. To stay in business he also produces local varieties that perform without use of fertilizers.

Unlike Sherif, Mama Adama has no relationship with rural development organizations and produces exclusively local varieties without any agrochemicals, thus reducing her production costs. Comptoir Agricole prefers buying seed from farmers and the research centre to minimize the risks of seed production. Although both enterprises have acquired a reputation among their clients, they differ fundamentally: while Mama Adama capitalizes on her social network to sell her seed, Comptoir Agricole relies on the social networks of young people who scout the area for quality seed at the time of harvest.

About 80% of the members of the Cereal and Potato Seed Producers' Union are women, eager to earn an income and feed their families. The Union has access to plenty of fertile land on the Fouta Plateau, as well as access to foundation seed and technical expertise through its management. As all are extension agents who have worked their entire lives with farmers, they have a vast network, deep respect and sound knowledge of farmers' needs.

El-Hadj Sow was inspired by a visit to France in the late 1980s, after which he developed strong institutional relations and still receives technical advice from research and extension. However, his client network is strong; he regularly visits his clients to support them and learn about their changing demands. He hardly sells to outside markets because of the lack of certification. The sustainability and reputation of Sow's seed enterprise rest on his entrepreneurial spirit, his solid management skills and his dedication to quality.

Besides the lack of subsidies and credit, rice seed producers face other challenges, especially scarce labour, which is in high demand during ploughing, weeding and harvest. Rice seed is produced at the same time as all the rest of the rice and during the growing season most households prefer to work first on their own farm, to assure their staple food supply. Paid labour becomes scarce and more expensive, making it difficult to expand seed enterprises.

Several seed enterprises in Guinea lack functional links with formal institutions, and make a business from selling quality seed of local varieties. Their networks and reputation are their major assets.

Various enterprises like the ones presented in this chapter survive without access to credit, which does not mean, however, that they would not benefit from more accessible and customer-friendly rural financial products and services. The government of Mali has done just that (Chapter 5).

Many publications on entrepreneurship in developing countries have argued that social networks are essential for entrepreneurial success. However, based on a study in Madagascar, Fafchamps and Minten (2002) stressed the need to distinguish different components of social networks, e.g. relations with other traders and with potential lenders increased transaction productivity, whereas extended family relations reduced it. Egbert (2009) came to the same conclusion, namely that African entrepreneurs in Tanzania financially support the extended families despite the trouble it causes for their business. However, the case of Mama Adama described in this chapter (Section 6.2) shows how extended family ties can have a long-lasting positive effect on seed trade. Social networks are indeed complex and context-specific; no doubt they can be supportive and functional, as well as parasitic.

References

Barry, B.M. (2006) *Diversité et dynamique des variétés locales de riz (*O. sativa *et* O. glaberrima*) en Guinée. Conséquences pour la conservation des ressources génétiques.* PhD thesis, Ecole Nationale Supérieure Agronomique de Rennes, Rennes, France.

Egbert, H. (2009) Business success through social networks? A comment on social networks and business success. *American Journal of Economics and Sociology* 68(3), 665–677.

Fafchamps, M. and Minten, B. (2002) Returns to social network capital among traders. *Oxford Economic Papers* 54(2), 173–206.

MAEF (2007a) *Politique nationale de développement agricole vision 2015. Vol. 1. Bilan-Diagnostic de la LPDA2 (1998–2005).* Ministère de l'Agriculture des Eaux et Forêts, Conakry.

MAEF (2007b) *Nouvelle vision de l'agriculture Guinéenne. Politique nationale de développement agricole vision 2015. Vol. 2. Orientations et axes stratégiques.* Ministère de l'Agriculture des Eaux et Forêts, Conakry.

Okry, F., Van Mele, P., Nuijten, E., Struik, P.C. and Mongbo, R.L. (2011) Organizational analysis of the seed sector of rice in Guinea: stakeholders, perception and institutional linkages. *Experimental Agriculture*, 47(1), 137–157.

Portères, R. (1966) Les noms des riz en Guinée. *Journal d'Agriculture Tropicale et de Botanique Appliqué* 13, 346 pp.

Richards, P. (1986) *Coping with Hunger: Hazard and Experiment in an African Rice-farming System.* Allen and Unwin, London.

SG2000 (2005) *Rapport Final 1996–2004.* Sasakawa Global 2000, Conakry.

SNPRV (2001) *Multiplication and Diffusion of Community-based Seeds of NERICA Rices Developed by WARDA.* Agence Nationale de Promotion et de Conseil Agricole, Conakry.

WARDA (2007) *Africa Rice Trends: Overview of Recent Developments in the Sub-Saharan African Rice Sector.* Africa Rice Centre, Cotonou.

7 The Gambia: Capturing the Media

J. David Reece, Daniel N. Dalohoun,
Essa Drammeh, Paul Van Mele and Saidu Bah

7.1 Introduction

7.1.1 Agriculture

West Africa's smallest country, The Gambia, is a 338 km long and narrow strip of land along both banks of the Gambia River. Except at the coast, it is entirely surrounded by Senegal. With a population of just 1.7 million people, agriculture drives its economic growth, but crop yields are low. With at least 90% of poor people working in agriculture, gains in productivity are vital for reducing poverty.

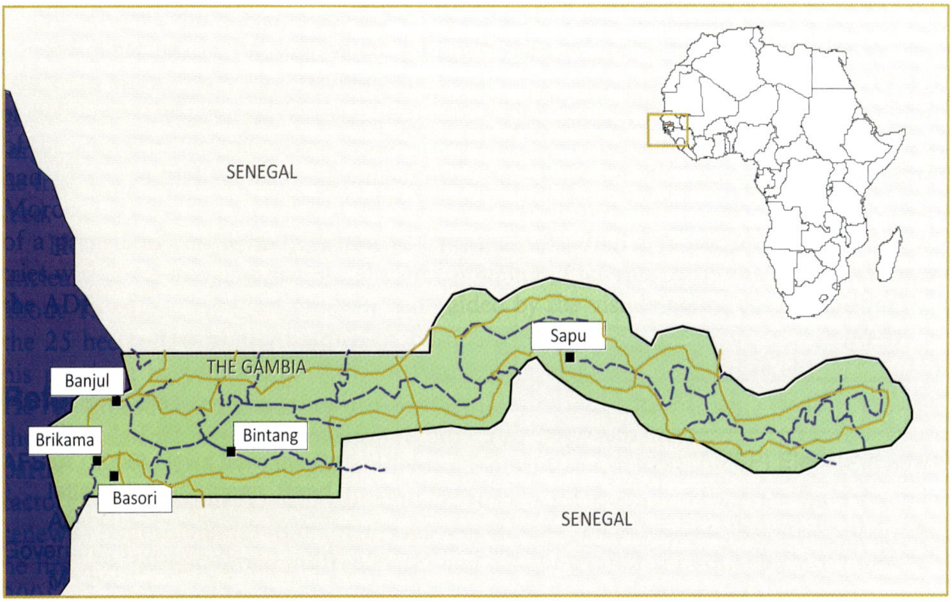

Gambian agriculture is vulnerable since the rainy season lasts for just 3 to 5 months and is unpredictable. About 52% of the land area of The Gambia (1 million hectares) is arable, but only about 60% of the arable land is farmed. Most cropland is owned by communities; individuals hold use rights but cannot sell the land. Family farms are often smaller than 1 hectare, and may be made up of several plots scattered about the uplands and lowlands of the village, so that each family has shares of the community's different types of land, allowing them to spread risk and grow more kinds of produce (Bentley, 1987).

Groundnut is the major cash crop, followed by sesame. Rice, millet, maize and sorghum are grown for home consumption. Rice is grown as household food but with increased production it is becoming a cash crop. Early millet was formerly grown to eat at home, but farmers are now also growing it as a cash crop, because of wild price fluctuations for groundnut.

The Gambia probably produces enough coarse grains to be self-sufficient in food, but much is exported since Gambians prefer eating rice, which is easier to prepare. Until recently, 80% of the rice eaten in The Gambia was imported, while coarse grains are exported, mainly to Senegal. Home-grown rice has a special flavour and is preferred for some local recipes, so it fetches a higher price than imported rice, but farm-gate prices are low at harvest time. Farmers generally do not have storage facilities or access to credit and so must accept postharvest prices.

After drying the rice and millet, women pound it with a mortar and pestle. Mills are still scarce in The Gambia.

7.1.2 History

Colonial policy during the mid 19th century encouraged men to concentrate on groundnut as a cash crop, leaving women responsible for growing rice for food in the lowlands (Nuijten, 2005).

Growth in agricultural production was made possible by industrial development in Senegal, which began to produce and export animal-drawn ploughs, seeders and weeders. The Senegalese bred superb draught animals (horses and donkeys) and traded them across the border. At the same time, the authorities in The Gambia helped to make these implements available to farmers, providing training and loaning the equipment through agricultural cooperatives. Most upland producers of groundnuts and cereals mechanized most of their operations, except for harvesting. Animal traction also allowed mechanized seeding and weeding for rice (Remington and Posner, 2000). As more land was put to the plough, agricultural output rose. Fertilizer was subsidized and easily available, so that yields remained reasonably high. In the mid-1980s, however, structural adjustment killed this policy and made way for private investment, which failed to move into the vacuum, leaving the farmers dangling.

From the 1970s through the 1990s the government invested in irrigated rice, to promote food security. Irrigation meant that smallholders could grow rice in the dry season, so men began pushing women out of the rice fields, taking over the land, the work and the increased profits (Carney, 1998). However, because irrigation relied on pumps and farmers were unwilling to pay for the fuel, the projects were soon abandoned. In 2006 the government resumed irrigation projects, managed with more local participation and using tidal irrigation instead of pumps.

7.1.3 Policy

Gambian public policy aims to modernize agriculture, even though the government has no overarching policy document. The creation of NARI in 1995 constituted a major policy intervention, representing the government's commitment to agricultural development. This was followed by the formation of a national extension authority, which gave the extension function enhanced importance and visibility, while additional resources were provided for agriculture-related training. Each major crop is served by at least one project, with others for food-processing and postharvest activities. Such projects provide funds for transport, making it possible for extension agents to work with farmers, despite the broader funding problems affecting this service. In practice, advisory services are available only to those producers who have personal contacts with them. Some producers also benefit from additional support services, such as the use of expensive equipment or assistance with tasks that require high levels of technical skill, and need to pay for such additional support.

Agricultural implements may be imported duty free, while the government tries to make fertilizer available at a reasonable cost, through projects. The recent increase in the price of fertilizer means that smallholders cannot afford it, unless it is provided through a project.

Just like Guinea (Chapter 6), The Gambia lacks formal rural financial institutions. This is in sharp contrast with Mali (Chapter 5) and Uganda (Chapter 10), which have both witnessed an evolution in financial products and services for agriculture.

Staff of NARI (National Agricultural Research Institute) saw poor access to quality seed as a nagging problem, particularly for groundnut and rice. Hence the government of The Gambia wrote a seed policy document in 2008 which has now been approved by the cabinet and will soon be implemented.

Although efforts are being made to harmonize seed laws across West Africa, at present there are no seed laws in force in The Gambia, and no system of seed certification. While NARI has the expertise to verify the quality of seed, this assistance is provided on request (and in some cases on payment of a fee) and so constitutes support to a seed producer's own efforts to provide quality assurance.

7.1.4 Rice seed sector

NARI. The only institute that produces and supplies foundation seed is the Seed Technology Unit at NARI. Under some circumstances it places contracts with farmers or farmer groups to produce foundation seed, under its supervision. It responds to demand, multiplying seed when there is a request from NGOs or from other government institutions. NGOs are the main distributors of new varieties.

NARI hosts the Gambian component of the African Rice Initiative, which distributes Nerica seed in eight African countries. In The Gambia, this is usually called 'the Nerica project', which provides farmers with seed, technical assistance and fertilizer when it is needed, but at market prices (high). While it began by using foundation seed from Guinea, it relies on farmers to multiply seed, which it then buys to distribute to other farmers. Table 7.1 shows the quantities of Nerica seed that have been produced in The Gambia since 2006.

Farmer seed producers. Up to the introduction of Nerica, few farmers specialized in seed. Farmers occasionally obtained seed of modern varieties and used it to grow grain, but, until Nerica was introduced, few produced seed for sale.

Table 7.1. Nerica foundation seed production (tonnes), The Gambia, 2006–2009. Source: NARI.

	2006	2007	2008	2009
Nerica 1	2.3	3.5	4.7	5.1
Nerica 2	0.1	0.9	1.9	2.6
Nerica 3	0.5	1.4	2.5	3.1
Nerica 4	1.4	2.6	3.8	9.2
Nerica 6	2.8	2.4	4.7	7.2
Nerica P105	2.3	5.5	10.0	25.5
Nerica P163	4.6	7.8	9.4	33.2
Total	14.0	24.0	36.9	85.8

7.1.5 Seed use and demand

Variety names. Farmers often use the same name for different varieties, or different names for apparently identical varieties (Nuijten, 2005). Likewise, the name Nerica is used to cover various improved varieties, with farmers seeming not to differentiate between them. NARI staff fear that farmers buying seed are unaware that there are different kinds of Nerica and other improved varieties for different contexts, and so may not look for varieties that match their needs and goals.

While farmers may be confused about modern varieties, female rice farmers are aware of the differences between different traditional varieties (Nuijten, 2005). The farmers Nuijten interviewed grew an average of three varieties to match different ecological niches, although some people had as many as 11 varieties. Women constantly look for new and better varieties, even when they already have varieties that meet their needs and preferences.

Farmers in The Gambia offer visitors from NARI some cool water and take the opportunity to ask what new varieties the scientists are developing.

Reasons to buy seed. In some cases farmers need to eat their entire harvest and cannot save any seed. Sometimes they find that the seed they are using has become so mixed that they would prefer to start again using cleaner seed. Another reason for buying seed is to obtain a promising new variety.

Grain market. Demand for rice seed is tied to demand for grain, which is high, since rice is the staple food of The Gambia. Gambians eat 107 kg of rice per capita each year, the third highest in West Africa (ADF, 2005). Rice is quick and easy to prepare, endearing it to busy urban consumers. Home-grown rice generally retails for more than imported rice. Rice is often eaten as porridge, and so the consumers don't care if it is milled mechanically or pounded with a mortar and pestle. Rather, they pay a premium price for the preferred flavour of local rice, while differences in milling quality are less important to them.

But it is not easy for farmers to get the high prices that they could get from local rice. Processing capacity is limited, and home-grown rice has little access to a market where the main distributors are still dominated by importers. Shops in Banjul offer a wide selection of imported rice but no home-grown rice at all. Farm households eat much of their own rice rather than selling it.

Nerica was introduced in the early 2000s with much publicity, while the initiative to distribute Nerica seed to farmers across the country benefited from high-level political support. The goal was clear: to enable The Gambia to grow more rice and cut its high bill for imports. NARI estimates that, considering the number of seed producers in the country, it will take some time to saturate the market. Most seed buyers are grain farmers who barely meet their food requirement and so are unlikely to sell seed to others.

Changes in seed use and demand through media. Before the arrival of Nerica, people thought of rice as an aquatic plant, not realizing that it could also grow on soils other than swamps and inland valleys. Upland Nerica varieties that the Nerica project grew along the main roadsides exposed the Gambians to a new and radical innovation everybody was eager to experiment with. People could appreciate the behaviour and performance of the new early maturing, high yielding rice varieties which also had high market value.

At the same time as Nerica varieties first became available, TV and radio broadcasts began to promote the virtues of the new rice varieties and explain the importance of quality seed. Farmers flooded the radio with questions about these varieties, where to get seed and how to grow it. The media attention given to seed, along with the arrival of upland Nerica, which is a new crop in The Gambia just as it is in Uganda (Chapter 10), has motivated farmers both to use quality seed and to cultivate modern varieties.

7.1.6 Media and agriculture

National radio and TV. Even prior to independence, rural radios in most African countries played a role in promoting agricultural innovation (Ilboudo, 2003). More recently, TV stations have also started to assume that role. From the moment of its commissioning in late 1995, Gambia Television has operated as a public service station in the tradition of the older Radio Gambia, which was established in 1962. Broadcasting in English and local languages, the Gambia Radio and Television Services (GRTS) continue to share information on agricultural innovations. In a country where 60% are illiterate (UNESCO estimate, data 2003), audio- and video-based communication is crucial.

When in 2000 GRTS radio staff read in a newspaper that researchers at AfricaRice (then called WARDA) had discovered new rice varieties that did wonders, they approached NARI for more information. NARI soon participated in live radio debates and started to invite GRTS staff to see and talk about what they were doing.

Also Gambia Television has been keen to keep abreast of agricultural developments. When NARI scientist Mustapha Ceesay approached GRTS with videos about rice seed health from Bangladesh, The Gambia became the first country to translate them into an African language, and regularly broadcast them (Box 7.1).

Box 7.1 Bangladeshi seed videos on TV.

On 30 April 2005, the National Agricultural Research Institute (NARI) invited three women from Pirang village to watch videos on rice seed health, produced with rural women in Bangladesh. Also the President of the National Farmers' Platform and representatives of the research and extension community attended. Based on the powerful feedback of the women, the people in the room immediately decided that these videos had to be translated into Mandinka, one of the main national languages.

Translating them took some time, but 2 years later, under the impulse of Mustapha Ceesay, a dynamic scientist at NARI, Gambian TV started to broadcast the Mandinka version of the rice seed videos.

Source: Van Mele *et al.*, 2010b.

The media attention triggered a particularly high interest among farmers, so GRTS started to allocate more time to Nerica. The national radio partnered with NARI to cover major events associated with Nerica. It regularly visited demonstration plots and fields where seed of the new rice varieties was being produced. The broadcasts sensitized farmers about the advantages of upland rice and stimulated development workers to assist growers, as well as prompting the government to boost support.

Rural private radio. The Gambian media being tightly controlled by the government, there are only a handful of privately owned newspapers and radio stations. Yiriwa Development Radio started in 2005 and a year later it launched a programme called 'Back to the Land', presented by Mr Omar Fofana. The programme was launched in Kanilai, the village of the president of The Gambia, where many farmers and farmer associations gathered to help the president on his rice farm. The president wanted to be a living model for his people by going 'Back to the Land'. He used to say: 'If you want, come and see what I am doing.'

The radio programme focused on food and staple crops, among which Nerica held the first place thanks to farmers' high interest in upland rice. The programme, which is presented in Mandinka, explained to rural people the advantages of Nerica, where to acquire the seeds and how to use them. Later on, it also informed its audience about rice video CDs that present good practices in rice production and processing, developed by AfricaRice and partners (Van Mele *et al.*, 2010a). It also directed listeners towards NARI and the Jambur Kafo (Section 7.3).

During the 1 hour per week programme people could present their farming activities, ask any question or get any information related to agriculture. In case the presenter could not answer some questions, he invited, in the next programme, agricultural specialists and knowledgeable farmers to intervene.

Unfortunately, this programme stopped due to lack of financing. Mr Omar Fofana left the radio and went to the West Coast Radio, where he continued with the programme for just 2 months with funds provided by the radio and Gambia Telecommunications Cellular Company Ltd (Gamcell). The programme costs Dalasi 12,000 ($480) per month. Mr Omar Fofana is still actively looking for sponsors to continue with this programme, which, he thinks, has helped many farmers and other workers concerned with agricultural development in the country.

Case studies. In what follows, we present three case studies. All three enterprises are located close to the NARI office, which is on the outskirts of Brikama, just south of the Banjul international airport. The first one describes the Gambia Horticultural Enterprises, a private company that trades rice, maize and groundnut seed, among other products and services. The second case is a community association, Jambur Kafo, which participated in varietal trials with researchers and later began producing Nerica rice seed and dry season vegetables. The third case describes the Jafaye Farm, a family business growing Nerica seed, among other farm enterprises. All have experimented with media to advance their business.

7.2 Gambia Horticultural Enterprises

7.2.1 History

The Gambia Horticultural Enterprises (GHE) was established in 1990, by Mr Momodou Ceesay, as a sole proprietorship registered company to sell agricultural inputs. Gradually it also started to deal in equipment and tools, and expanded its services. GHE is located a few kilometres from the capital city of Banjul and has become the country's largest input enterprise.

In his early years, Mr Ceesay was an agronomist within the Department of Agriculture. Later he became the general manager of Citroproducts Gambia Ltd, a quasi-parastatal agribusiness focusing on lime production and processing. In this position he learned about market demands for various horticultural products, agricultural inputs, equipment and other tools.

In almost 20 years GHE has grown from a small office in the owner's home to much larger premises (a large building housing a shop selling agricultural supplies, another shop selling produce for consumption, warehouses for each shop as well as offices for three managers and their support staff), which the company bought after a period of renting. Initially, GHE sold only vegetable seeds, which it imported from the Netherlands (Table 7.2). Higher yields after using quality seed increased demand for it so GHE imported more. From 1993, GHE also started to import agricultural equipment and agrochemicals. Farmers were also interested in seed of field crops such as maize, groundnuts, rice, sorghum and millet. To respond to these requests, GHE buys from local producers, including the Jambur Kafo discussed later in this chapter.

GHE's growth, then, has been driven by market demand, with neighbouring countries (Senegal, Guinea Bissau and Sierra Leone) mainly requiring vegetable seed while most sales of seed for field crops have been in the home market. While the high interest in Nerica varieties explains the importance of rice seed, maize seed demand is increasing from farmers growing maize to supply the peri-urban poultry industry. Although seed sales have increased over the years, they now provide only 30% of the company's revenue. Other cash earning activities include the growing and marketing of fresh fruits and vegetables for local and European markets, which it started in 1995.

Table 7.2. Seed sales, Gambia Horticultural Enterprises. Source: GHE.

	1990	1993	1997	2000	2005	2009
	Importance (%) of seed sales in total company's turnover					
Seed sales	100	80	70	50	40	30
	Seed quantities sold (tonnes)					
Vegetables	0.5	1.0	2.0	2.0	3.0	5.0
Field crops	–	1.0	2.0	3.0	20.0	50.0
	Share (%) of individual field crops in the total sales of field crops					
Rice	NA	NA	NA	NA	50	50
Groundnut	NA	NA	NA	NA	20	10
Maize	NA	NA	NA	NA	20	30
Sorghum and millet	NA	NA	NA	NA	10	10

7.2.2 Structure

Management. Mr Ceesay is the managing director. The enterprise has over 70 employees, including a secretary and personnel in charge of production and marketing. Mr Ceesay's goal is to provide a one-stop shop for farmers. To achieve this, the company has established a garden centre, a food centre, an equipment centre and pest control services. The garden centre sells seed of vegetables, groundnuts, rice and maize, along with mango and papaya seedlings. It also deals in simple garden tools and agrochemicals.

Mr Ceesay believes that the introduction of Nerica varieties drastically boosted investment in agriculture, which was further stimulated by the president himself taking the lead in producing rice and encouraging others to invest in the sector. The recent introduction of milling equipment into the Gambian market has triggered further investment and development along the rice value chain. GHE has already sold 25 milling machines to NGOs, government agencies, individual farmers and projects.

Land and infrastructure. GHE headquarters are located along the Banjul–Serrekunda Highway within the Greater Banjul Area: a strategic location for marketing, much like Faso Kaba in Mali (Section 5.2). There is also a sales and production division. Until recently, the company owned only a few hectares that were mainly used to test batches of seed before marketing them. Lately, the company has acquired over 100 hectares of farmland, and plans to use this to produce some seed itself rather than relying entirely on external suppliers. It has also recently opened a food-processing centre, which will enable it to expand the range of processed foods that it sells.

Links. Mr Ceesay is president of the Gambia Agrochemicals and Seed Trade Association (GASTA), for which he provides secretarial facilities. Until 2009, he was also the chairman of the Board of Directors of NARI. If need be, GHE sponsors NARI to conduct advanced yield trials of promising new varieties, including those that it is considering importing from Europe but which are as yet untested in The Gambia.

All along, Mr Ceesay has maintained good links with research, other workers, the national farmers' platform and the media. GHE has also developed good relations with commercial banks and business partners abroad.

Quality control. GHE will only buy seed from producers whom NARI knows to be established producers of good quality seed. Before accepting a new supplier, GHE with the support of NARI researchers (paid by GHE) tests each variety from a new producer to ensure that its performance is satisfactory. Only after completing this will GHE retail seed from this supplier. These tests are generally undertaken once for each variety from each supplier: after this, GHE will normally trust its suppliers so that further quality control is not needed, although samples of packaged seed are tested for moisture content and germination rate at the beginning of each sowing season. Furthermore, all customer complaints are investigated: GHE first verifies that the customer followed the recommended cultural practices for the variety grown, and then again tests seed from the producer responsible to establish the reason for the problem. All these trials take place on GHE's own land.

7.2.3 Cash flow

While heading Citroproducts Gambia Ltd, Mr Ceesay felt that there was a dire need for a private agribusiness company; there was none in The Gambia at that time. He also thought that investing in such a business would be profitable. Mr Ceesay prepared a project document and took it to the bank for financing, using his home as collateral, and obtained a 3-year loan at a 26% interest rate, which he readily paid off. With his track record as a reliable customer, he has since been able to borrow at lower interest rates. In 2000 he had access to credit at 22% interest, and in 2010 he could borrow at 20%.

GHE relies primarily on its own funds but still needs to borrow sometimes, and is able to do so since it enjoys good relations with its bank. In general, GHE pays cash for seed and also sells it for cash. The enterprise buys seed at harvest, applies seed treatments to protect it from soil insects and stores it until it can be sold.

7.2.4 Marketing

Just a few weeks before sowing time, the seed is packaged in smaller bags of 0.5 kg, 1 kg, 2 kg, 5 kg, 10 kg, 25 kg and 50 kg to meet the varied demands of GHE's diverse clients. Up until 2009, the most important of these clients was a combination of international organizations, projects and government agencies (Table 7.3). They bought seeds of all crops and supplied them to poor farmers in order to demonstrate the advantages of using improved crop varieties. However, GHE thinks that this

Table 7.3. Clients of Gambia Horticultural Enterprises.

	1990	2000	2009	2015 (predicted)
Government agencies	2	2	1	3
Individual farmers	4	4	2	2
Companies/farmers' organizations	3	3	3	1
International organizations and projects	1	1	4	4

Ranking assessment by senior management of seed enterprise, 1 being the most important.

kind of assistance will gradually come to an end. From 2015 and onward, individual farmers and their organizations (in diverse forms) are expected to be the main customers for the seed business, since by then they would have already experienced the benefits that result from the use of good quality seed. In particular, sales to individual farmers are expected to increase considerably, although their demand will still be less than that exerted by the emergent or growing agricultural business enterprises. However, GHE foresees that government efforts to improve agricultural production will continue, and that these will include the provision of seed supplies for farmers. Government agencies will thus be the third most important customers of the seed business, while international organizations and projects will fall into fourth place.

From the very beginning GHE invested in advertising its products and services, especially over the radio (in English, Wolof and Mandinka) and in the print media. Before the sowing season, Mr Ceesay negotiates with GTRS, which has a weekly educational radio programme on diverse themes including agriculture. As the first company of its kind, GHE obtained good deals to regularly advertise during and after this programme.

GHE has diversified its portfolio by adding seeds of new crops, as shown in Table 7.2. So, although the company is well known after 20 years of activity, it continues to advertise, partly in order to build markets for its newer products and partly to maintain its position in the face of competition from the other companies that have more recently entered the seed market. Advertising intensifies at the start of the cropping season. The enterprise also has its own website (www.gamhort.gm).

Mr Ceesay participates in national and international agricultural fairs to market the company's products and to keep abreast of potential new varieties for The Gambia and surrounding countries.

GHE acknowledges the existence of about 11 other companies of its kind but fears no competition. With its long experience, its diversified range of products and services, the quality and regular availability (avoiding stock shortages) of products and services, with appropriate advertisement, GHE has many advantages over its competitors.

While NARI provides support and assistance for all companies in the sector, Mr Ceesay has been the NARI chair for almost a decade, and has cultivated this relationship as part of his marketing and customer service strategy. Under negotiated conditions that include the payment of appropriate service charges, NARI provides specialized services for GHE and its clients when they face a particular challenge. For example, if the crops planted by one of his customers suffered attack by insects or diseases, GHE invites researchers from NARI to examine the case and search for appropriate solutions. Furthermore, if a customer complains that seed has not germinated properly, GHE invites NARI to investigate in order to find out whether the problem results from any deficiencies in the seed or from the customer sowing the seed incorrectly.

7.3 Jambur Kafo

7.3.1 History

Kafo means 'community association' in Mandinga. The Kafo in the village of Jambur, on the fringes of Banjul and near the NARI office, was founded in 1994. It owns 16 hectares that are farmed communally, while its members also have their own individual landholdings. The Kafo uses the communal land to grow 50 kg of rice seed for each member, with any surplus sold to boost the Kafo's funds.

The Jambur Kafo has a long history of promoting modern rice varieties. Long before Nerica came, the Kafo took part in participatory varietal selection (PVS) trials of modern varieties and multiplied seed provided by NARI. The seed that they produced was given to individual members of the Kafo and to farmers in neighbouring villages, who grew their own seed the following season. Some of the harvest was used as grain and eaten at Kafo activities; the Kafo was learning to multiply seed, but not to sell it.

Before starting to grow Nerica, the Kafo (comprised mainly of women) had undertaken various commercial activities including soap making and tie-and-dye, although they later abandoned these activities to concentrate on producing rice seed in the rainy season. They still grow vegetables in the dry season.

After harvesting the rice seed crop, individual members grow vegetables in the dry season and pay part of their earnings from sales back to the cooperative.

They began producing Nerica seed in 2002, a year before the start of the Nerica project. The seeds were originally given to them as part of the PVS (with the Ministry of Agriculture and AfricaRice). This meant that, when the Nerica project began, they had plenty of seed and so the project asked them to sell their seed to the project and to other farmers. This was probably the first time that anyone had sold upland rice seed in The Gambia.

The Kafo began seed production in 2002 with the line WAB 450-1-B-P-163-4-1, which the Kafo members named white Nerica or P163 (Table 7.4). The following year, they got another Nerica variety from the Nerica project at NARI. It was Nerica 3 but they called it Conakry because the Nerica project got it from Guinea. The Kafo first tested the white Nerica against other modern varieties that they had earlier helped to disseminate, as well as their own local variety. During a drought in 2003 only white Nerica thrived, while all the other varieties failed to mature or suffered reduced yield. After that most seed buyers asked for white Nerica and would only buy a different variety if all the white Nerica had been sold.

NARI later gave the Kafo more lines to evaluate, so from 2004 they added Nerica 4 and WAB 56-50. In 2006, the Kafo inserted another variety, Nerica 1, into its fields. While most customers demand white Nerica, the Kafo members observed that in more favourable years the other Nerica lines performed well, and so they decided to keep cultivating them.

As seed demand increased and members gained experience in seed production through training from NARI, from 2005 onward members began giving greater priority to producing seed on their own individual landholdings. Most of them only started working on the Kafo farm after finishing most of the work on their own farms. Crops on the communal land thus started suffering from poor care, mostly

Table 7.4. Rice seed produced (tonnes), Jambur Kafo.

	2002	2003	2004	2005	2006	2007	2008	2009
Nerica 1	–	–	–	–	0.8	1.3	0.9	0.9
Nerica 3 (Conakry)	–	2.1	3.2	1.8	1.0	1.5	1.5	–
Nerica 4	–	–	1.8	3.1	1.1	1.2	0.3	–
P163 (white Nerica)	0.9	21.2	36.6	22.0	4.3	2.8	1.9	1.2
WAB 56-50	–	–	1.4	3.5	2.1	1.3	1.2	–
Total production	0.9	23.3	43.0	30.4	9.3	8.1	5.8	2.1
Total area cultivated (ha)	0.5	6.5	13.5	16.0	16.0	10.0	10.0	5.0
Changes (%) in total production from year to year	–	2,589	85	–29	–69	–13	–28	–64
Key reasons for change	Start of Nerica project in 2003		All members of the cooperative were trained in seed production		Members increasingly focused on individual seed production (data not included in this table) rather than on cooperative fields to respond to increased demand for seed			

from delays in seed production activities. At their general meeting at the end of 2009, the Kafo members decided that, from 2010 onward, they would concentrate on producing just two varieties, Nerica 1 and white Nerica (P163), and would produce seed on only 5 hectares. The remaining 11 hectares of the Kafo's land was lent to individual members so that the Kafo does not lose control of it, and will be taken back whenever the Kafo needs it again.

Another reason for declining yields is that since 2002/2003 the Kafo has never renewed the foundation seed that it uses. It will, however, use fresh foundation seed for the two retained varieties in 2010. Observing the development of the seed crops on their own farms and feedback from clients 'that the seed no longer worked' also contributed to this decision.

7.3.2 Structure

The association. The Jambur Kafo is an association of villagers who all have their own land and so farm as individuals as well as collectively with the Kafo. Since 2000 the Kafo has had a steady membership of 200: 180 women and 20 men. The Kafo does not pay its members for their labour, although they enjoy a communal meal on workdays.

In the dry season, individual members grow vegetables on the Kafo's land and pay part of their earnings from sales to the Kafo. This income helps to maintain the garden, especially the wells and fences. Since there is no rice farming in the dry season, it was easy to diversify into vegetable gardening. In principle this could provide finance for the rice-planting season, but the Kafo rarely transfers money between the vegetable and rice activities. The vegetables are also used to feed members when they are working on the communal lands or are shared with visitors during the field days.

Management skills (working in groups and through committees) learned in the rice field also work just as well in the vegetable garden. The two enterprises have the same type of committees and the same people work in different roles within the different enterprises. For example, there is a committee that is responsible for ploughing. They go and look for a tractor, hire it and supervise the ploughing. There are similar committees responsible for the other operations.

Land, infrastructure and equipment. When the Kafo began it did not have any land and so approached a landowning clan, and was given land, free and simple.

Fairly soon after starting to produce Nerica in Jambur, the Nerica project arranged for a visit from the president's wife. She was so impressed that she gave the Kafo 5000 Dalasi ($200) and 25 bags (1.25 tonnes) of fertilizer. As a result of this visit the President himself later donated 1000 bags of cement and some corrugated sheet metal to help them build a seed storehouse. The Kafo members began building the storehouse in 2004, providing their own sand, gravel and labour. But they gradually saw that they couldn't finish the building on their own, and so the Nerica project eventually helped with sand and roofing materials, enabling the Kafo to complete the large storehouse in 2007; it includes a fenced area where they dry rice. They cover it with tarpaulin for drying since they do not have a cement drying floor.

The Gambian President donated cement and corrugated sheet metal for the Jambur Kafo seed storehouse.

In 2005 the Nerica project sold the Kafo a seeder and a donkey-drawn weeder on credit. This loan has now been repaid from sales of rice seed. In 2009 the project donated a power tiller, to expand production into nearby lowlands. Some members of the Kafo operate the power tiller and threshing machine, but do other farm chores when the machinery is not being used. Since these machines belong to the Kafo, they are used primarily for work on the Kafo's land. It is only when work on the collective land has been finished that the machines are hired out to work on the plots of individuals, with the machine operators receiving a share of the money raised in this way.

Evolution of technology. Since the Kafo first started growing Nerica its members have always planted in rows with a donkey-drawn seeder, rather than broadcasting seed. They also use donkeys to pull the weeder and then they hoe up the weeds within the row. And they have continued using mineral fertilizer.

Training. Training was provided by the Department of Agricultural Services, covering fertilizer and crop management. The Nerica project gave training in operating the power tiller and threshing machine. The Kafo thought the training was too short, so a technician came and worked with them for a week, and they learned how to use the machines by working under supervision. They believe that they now need training on the rate and timing of herbicide application. They do not use herbicide, even though they know that it is available and works well for early weed control, because they don't know about safety measures, when to apply it or the dosage. They believe that knowing this would enhance their ability to control weeds and so increase production.

Quality control. The Kafo members do not test seed for moisture content and germination rate, but stress the purity of each of their varieties. They take care to plant one variety at a time, using only land that had previously been used for that same variety (to avoid volunteers creating impurity). At flowering and at harvest time the women remove off-types (rogue) to ensure purity. Few other seed producers are able to undertake these time-consuming, demanding practices.

After harvesting, each variety is dried and winnowed separately. Sometimes a variety will be assigned to one group of people to process at a time. Thus each variety will be threshed separately. When changing varieties the first two bags threshed after changing variety are used only as grain, to avoid mixing. The store is cleaned thoroughly before storing the seeds. And they generally dress the seed that they are going to plant, to help keep it disease-free.

Linkages. The Kafo has a close relationship with NARI, which has agreed to give them new varieties and technical assistance, and has ties with TV and radio stations. The Kafo also enjoys good working relations with other seed producers in the region, and so was able to agree to a seed price increase when the cost of fertilizer increased. They supplied seeds and technical assistance to Jafaye Farm (Section 7.4) and continue to cooperate with its owner.

Collaboration. For large orders, the Kafo collaborated with other kafos and large-scale seed producers to meet it. They are expecting another large order in 2010, but do not believe that any of the other kafos can produce to their own high standards of purity and so are unsure how to meet the order and keep their customers' trust. One option is to use seed produced by individual members, but that may not be enough.

Competition. The Kafo at Jambur competes for sales with other kafos and with large-scale farmers. However, they believe that they retain customers because of the purity of their seeds and because everyone knows that they are well-established and reliable seed producers.

7.3.3 Cash flow

Evolving input costs. Seed prices have changed in line with the Kafo's production costs, rather than responding to changes in market demand. When they started producing Nerica they charged 15–20 Dalasi ($0.60–0.80) per kilo, but rising fertilizer prices forced them to charge more. In 2009 they charged 35–40 Dalasi per kilo ($1.40–$1.60). This price increase was agreed with other producers in the region to avoid undercutting each other. They actually did undercut each other a little, but less than if they had not agreed on the price increase.

Operating capital. The Kafo does not need external financing every year, partly because it does not pay for labour and it has limited costs (mainly for fertilizer and fuel for the machinery). The Kafo built up operating capital by growing and selling Nerica seed. They are unwilling to use credit, following one experience. They hated paying interest at a high rate and have never borrowed since. Even so, they feel that they now need credit to buy a tractor. In 2009 they applied for a bank loan, which the bank approved, but it has still not released the funds.

Credit. The Kafo does not sell seeds on credit either, because of another bad experience: they gave seeds on credit to the Soil and Water Management Unit of the Department of Agricultural Services (DAS), which distributed them to groups of farmers who had built water retention dykes. These farmers agreed to repay the seed in kind after harvest, but the seed they gave back was so mixed it could not be sold. Those who had promised to repay in cash simply reneged. Since then, sales to outsiders have been for cash only. However, people they know and trust (Kafo members and farmers from neighbouring villages) are allowed to buy seed on credit and pay later in cash.

7.3.4 Marketing

Public profile. The Kafo's seed business attracted huge interest thanks to its partnership with NARI and the mass media. Right from the start the Kafo enjoyed a high public profile because NARI arranged for important people to visit the rice fields at Jambur and ensured that these visits received national publicity. NARI used to invite media, and expected most of them to cover such an event. In addition, a throng of media always followed prominent visitors like government officials. Seed customers were attracted from far away.

Clients. The Kafo has many customers. So far, small individual farmers like themselves buy 75% of the seed that they sell (Table 7.5). However, this figure includes the seed sold to agro-dealers who present themselves as individual farmers when coming to buy. Kafo members speculated that they feared that the Kafo would increase its seed price if it knew that they were enterprises. Until we discussed this question with the Kafo members, they had no idea that they had sold seed to any enterprise or agro-dealer such as GHE.

Table 7.5. Clients of Jambur Kafo.

	2005	2009	2015 (predicted)
Kafo members	–	–	–
Individual farmers	1	1	3
Projects	2	2	1
NGOs	3	3	2
Agro-dealers	–	–	4

Ranking assessment by senior management of seed enterprise, 1 being the most important.

The next most important clients were projects. Some projects have also bought seed to distribute to the farmers they serve, such as the Africa Emergency Locust Project (AELP) and the Special Programme for Food Security. In 2007, the Red Cross and Voluntary Services Overseas (VSO) bought seed, while the AELP came to buy in 2008. Since then, other projects and NGOs have bought seed. The Kafo thinks that in future such projects will become more important than NGOs and individual farmers.

Radio and TV. During the Nerica project the Kafo members noticed that, whenever ministers visited, they always came with a TV crew and that the publicity helped to attract customers from far away. So they took the initiative to build relationships with TV and radio stations, developing skills at working with the media. Once or twice they paid the cost for TV stations to come and visit their fields. They were providing programme content rather than paying for advertising but needed to meet the travel and accommodation expenses of all the people involved. Even though they paid expenses, getting a TV crew to come was easier said than done. Sometimes they arranged a date only to have big news events take place elsewhere, and the TV staff covered those instead. And once the TV crew came, but without a TV camera, so the Kafo hired one.

They also need to arrange events for the TV crews to film. To organize a field day, they write letters of invitation to many farmers, kafos, NGOs and others and pay to advertise the occasion on the radio. They have to cook for their visitors; for one field day they slaughtered a bull. After lunch, the Kafo members show their visitors around the fields. Despite the cost, they know that this kind of promotion pays. Broadcasting field days allows the Kafo to show the quality of their rice fields to potential customers and so sell seed to a far wider area than could normally be reached by a community-based seed enterprise. Some customers said that they were looking for Nerica seed and didn't know where to get it until they heard on the radio or TV about the seeds in Jambur. Others only decided to plant Nerica after admiring TV images of the fields at Jambur. There were instances when would-be customers were still telephoning or even travelling long distances to visit them, even after they had sold all their seed. So the Kafo members felt that media worked for them.

In 2005 and 2006, when they produced a lot of seed, they used Kafo funds to pay Radio Gambia to carry announcements that they had Nerica seed for sale. Mr Bakary Fatty, a well-known radio journalist, comes from Jambur. Knowing him helps the Kafo contact the national media. Sometimes they ask him to facilitate the event, and sometimes he joins the media crew. But even with his help they still have to pay costs.

As well as the government-owned station, two private radio stations serve the Jambur area, Yiriwa Station in Brikama and Hill Top Station in Tabokoto. Both stations carry agricultural programmes. Omar Bojang, the secretary of the Kafo, was often invited to speak on the programme 'Back to the Land' and each time he would tell the audience that Jambur had seeds for sale. It was free to appear on 'Back to the Land'. So Mr Bojang used both of these stations to publicize the seed from 2007 to 2009.

Although radio has been used by government and private seed enterprises in Nigeria (Sections 4.3.4 and 4.5.4), seed dealers, farmer associations and cooperatives in Mali (Sections 5.2.4, 5.3.4 and 5.5.1), the Jambur Kafo also mobilized the TV to show what difference quality seed could make, to expand its customer base.

But the Kafo understands that media alone are insufficient to keep its customers as long as it cannot continue providing top quality seed. The renewal of its foundation seed is long overdue.

Mr Bojang, secretary of the Jambur Kafo, scares birds with his slingshot. He is an experienced rice farmer who regularly features in radio talk shows.

7.4 Jafaye Farm

7.4.1 History

Jafaye Farm is in Busambala, on the outskirts of Banjul. It is owned by Mr Alagie Dembo, who runs it with his employees. The business was officially registered in 2006.

Mr Dembo began farming in 2004 with cashew, groundnuts and maize after returning from living in Europe. But then he saw the President on TV talking about Nerica, and this inspired him to plant rice. He saw on TV that Nerica seeds were available in Jambur, and so went to a field day to see the new rice. He was amazed by the fields and this motivated him to move into rice farming, so he bought seed and became a producer. Initially he wanted to grow grain and even bought a mill, but Dr Kunjo, then the coordinator of the Nerica project, advised him to concentrate on producing and selling seed. Dr Kunjo explained that Nerica seeds were scarce and in high demand, and so if a relatively large farmer like Mr Dembo were to become a seed producer then farmers throughout the country would benefit. By the time he began to produce rice seed in 2005, the Kafo at Jambur had already been producing for about 2 years and so was able to give him advice, and seeds.

By 2007, he was planting 30 hectares to rice seed, yielding 800 bags of seed (with each bag weighing 50 kg). In 2008 he planted 25 hectares to rice seed and used the other 5 hectares to grow maize. There are several large poultry ranches nearby, so he tried growing maize as chicken feed.

Table 7.6. Rice seed production, Jafaye Farm.

	2005	2006	2007	2008	2009
	Quantity of seed produced (tonnes)				
P163	3.5	6.0	25.0	10.0	–
Nerica 4	2.0	3.4	15.0	5.4	–
Total	5.5	9.4	40.0	15.4	–
	Cultivated area (hectares)				
P163	6	8	20	17	–
Nerica 4	3	4	10	8	–
Total	9	12	30	25	–
	Seed production per unit area (tonnes per hectare)				
P163	0.6	0.8	1.3	0.6	–
Nerica 4	0.7	0.9	1.5	0.7	–

However, for reasons that are unclear, in 2008 he produced only about 15 tonnes of rice seed on the 25 hectares, less than half of his previous harvest (Table 7.6). He recognized that the quality of the 'foundation' seed he sowed partly accounted for the unsatisfactory yields in 2008. He has not renewed the foundation seed that he first got from Jambur Kafo in 2005. Jambur Kafo, which will renew its foundation seed in 2010, also reported a decline in quality of the foundation seed it has been recycling since 2002/2003.

By 2009, the land around Mr Dembo's farm had become so urbanized it was no longer suitable for agriculture, so at the end of 2009 he sold his rice field but not his young cashew grove. Then he bought another field, in Basori, 5 km away. The new farm is larger, so the area under rice will increase by 2010. He continues to receive training and advice from NARI, mainly on farm management and crop production.

7.4.2 Structure

Management. Mr Dembo is interested in getting good seed to farmers and in regenerating the land, rather than simply maximizing profits. He is keen to improve his farm and mechanize it, to reduce drudgery and to improve timeliness. He hopes to use more fertilizer to keep yields high.

He has ten permanent staff members who have worked for him since he first planted rice. Mr Dembo has an experienced farm manager with more than 10 years' experience working at a large Lebanese-owned farm producing vegetables and fruit trees. At periods of peak labour demand, up to 60 day labourers may be engaged. His tractor driver only works the days he is needed.

Land, infrastructure and equipment. Mr Dembo owns his own farmland. His old and new rice farms were both rain-fed. He has exclusive access to a seed store on the main road; it was hardly ever used, so now he borrows it. The only machines that he uses are the tractor, the rice mill and the animal-drawn seeder and weeder.

How the whole enterprise fits together. The new plot, of 100 hectares, will be used to produce rice seed, timber from gmalina trees and mangoes. Mr Dembo has just planted the trees, and will grow rice around them. Weeding the rice will keep the land clear, so he will not have to weed the trees. The rice will earn money until the mango trees mature. Meanwhile the cashews are still providing cash to invest in rice.

Mangoes are in high demand and so are likely to be profitable. He is planting the gmalina trees to reverse deforestation. Once the trees are mature they will produce timber and will regenerate after the timber has been harvested.

Evolution of technology use. Mr Dembo began by using donkey- or horse-drawn implements for ploughing, weeding and seeding. In 2007 he bought a tractor for ploughing. Depending on what is most appropriate for each field, either men harvest with a sickle or women use a knife for panicle-picking. For threshing, he borrows a machine from the Department of Agricultural Services.

Mr Dembo owns a mill but has never used it (apart from testing it) since he produces seed: he had already bought it when he was advised to be a seed producer, and so has had no use for it and 'does not have time' to operate a milling business for other farmers. He may use it in the future to become a rice (grain) producer since he believes that the seed market is nearly saturated. But he said that seed sells for more than grain, so shifting from seed to grain production would cut his net income by perhaps 70%. He explained that, after selling a single bag of rice to be used as seed, he would have enough money to buy a bag or two of imported clean rice to feed his family, but he would have to mill three bags of paddy to get a single bag of clean rice.

Mr Dembo has applied fertilizer at the recommended levels since he began growing rice, and has enjoyed an assured supply of fertilizer (at market price) through the Nerica project. However, he has never used herbicide, since he is not sure how. And he pays attention to bird-scaring.

Quality control. Mr Dembo sows with seed that he has himself saved, which he ensures is pure and has a high germination rate. His knowledge of seed management is based on advice from the Jambur Kafo, as well as general training in farming from NARI. All his seed is descended from the first seed he bought from the Kafo at Jambur. He plants different varieties apart with a gap between them. He eliminates as much grass as possible, so that his harvest is not contaminated by grass seed.

Mr Dembo harvests at 80% maturity so that the crops do not get dry and shatter at harvest. Like the Jambur Kafo, he practises roguing. He then dries the harvested grain so that it does not get mouldy, and threshes it mechanically. He takes care to dry the different varieties separately, for 4 to 5 days. He winnows the rice and dries it again for a day or two before bagging it.

Before storing his seed, Mr Dembo sends a sample to NARI to test for germination and purity. NARI thus assists him to maintain high standards of seed quality, although it does not have any responsibility for regulating seed production or for maintaining quality. NARI also provides seed-dressing fungicide (free, given as part of the Nerica project), which he applies to the seed. He does not use any special kind of packaging.

Linkages. Mr Dembo's closest link is with NARI, whose Nerica project guarantees his timely access to fertilizer at market price, conducts germination tests and gives him seed-dressing chemicals. Another important link is with the Department of Agricultural Services, which provides technical advice and lends Mr Dembo the threshing machine.

Mr Dembo also enjoys strong relationships with the other rice seed producers in his region, Kombo. He is the president of a regional association of local rice production kafos, including the one at Jambur. The association's main achievement was to agree on new seed prices (Section 7.3.3) after fertilizer prices increased. He has a stronger partnership with the Jambur Kafo than with the others. When he began farming, Jambur sold him seeds and gave him the technical advice that he needed. He also discusses and agrees on prices with Jambur on a regular basis and so regards them as partners rather than competitors.

7.4.3 Cash flow

Jafaye Farm stores the rice seed it produces from harvest time (October) until farmers start buying immediately before the next sowing season (April). The delay in receiving money places a strain on Mr Dembo's cash flow. However, he has other activities that earn money and so he does not need loans. The main reason for official registration was to open a company bank account, but Mr Dembo has not used it as a source of credit. Nor does the business receive any supplies on credit. He pays cash for fertilizer, fuel and other inputs. He sells seed for cash.

7.4.4 Marketing

Evolution of sales. Unsatisfied demand is lower than in the early years, since now there are more seed producers. Mr Dembo may respond by using his mill to process some of his harvest as grain.

At the same time, other farmers are clamouring to use his mill, particularly since the price for local rice is higher than imported rice. Pounding of rice by hand is unpleasant and hard work, always done by women, who are eager to switch to machines. Since rice mills are scarce in this part of The Gambia, a rice mill could charge high fees.

Evolution of seed prices. Mr Dembo began by charging 25 Dalasi ($1) per kilo, but fertilizer prices increased sharply. He and Jambur agreed to increase the price to 40 Dalasi ($1.60). He charges NGOs an additional 5 Dalasi ($0.20) per kilo because they can afford to pay more than farmers. Some people would pay much more. Once a businessman bought seed from him and some other producers at 25 Dalasi and then, once Dembo and Jambur had sold all their seeds, he resold them at 100 Dalasi ($4) per kilo. He had repacked them in plastic bags and sold them through an agro-dealer, giving the impression of high quality. Mr Dembo, however, simply weighs the seed and sells it in large bags. He said that he was not interested in maximizing income, like the businessman, but wanted to get seed to the farmers. 'Our objectives are different.'

Evolution of type and importance of clients for rice seed. Individual farmers and NGOs buy seed from Mr Dembo (Box 7.2). About 75% of his seed is sold to individuals, a proportion which has remained reasonably constant, while sales to NGOs fluctuate.

Box 7.2 God will reward him.

Alagie Mori Kebba Touray, from Faraba Banta village, farms just under 2 hectares of rice, as well as cashew, groundnut and millet. In 2006 he heard about a new and high yielding rice variety, Nerica, and decided to grow some himself. A friend told him that Mr Dembo was a large farmer who grew Nerica and sold seed. Mr Touray called him immediately and arranged to go to his house to ask him about Nerica. Mr Dembo recommended Nerica, but had no more seed available, and so advised him to try either Jambur or NARI.

Nerica seeds were scarce and demand was high, so suppliers were only selling small amounts to each customer. Mr Touray was keen to begin production, rather than spending a year or two multiplying seed, and so he bought Nerica from several suppliers: from the Kafo at Jambur; from NARI at the nearby village of Brikama; from Bintang and even from Sapu, a 4-hour drive away. He went wherever they had seed.

When asked what he thought of the quality of the seed that he had bought, he seemed puzzled. 'It was all Nerica, so it was excellent.' He clarified that when he bought the seed he wasn't looking for quality – in fact one batch didn't even germinate – he just wanted some kind of Nerica seed.

Although he calls all the seed he bought 'Nerica', he knows that there may be differences between the seed from each source. So he planted each batch in a different place and observed their performance. He is careful to produce pure and high quality seed: he checks his fields for off-types; examines the seed and sorts it, using only the best rice for sowing and for selling as seed, while the rest is eaten as grain. In this way he has produced his own seed and maintained each type separately ever since the one and only time that he bought seed. His methods of seed production are based partly on his own ideas, partly on advice from Mr Dembo.

He sells seed to farmers in the region. He sells all the seed he has available, and would be able to sell more if he could produce more. He tells his customers how he grows seeds, how he plants and how he applies fertilizer. In short, he passes on everything that he was told by Mr Dembo. He tells each customer they are welcome to come back for advice later.

Mr Touray produces more than enough rice to feed his family. He also sells Nerica seed for 25 Dalasi ($1) per kg, and his earnings from selling seed may or may not cover his production costs. He believes that one day God will reward him for helping his neighbours to grow Nerica.

Farmers who buy from him are generally new faces who buy only once, although some return, usually because they have lost their seed or found that it has become highly mixed. He gives customers technical information by answering their questions. He encourages repeat business by offering a bonus (more seed at no extra cost) to farmers who buy more than once. He believes that he retains customers because of the quality of his seed and his accessible location. His farm is near the highway and so customers can reach him easily, while getting to Jambur is more difficult.

Filming field days. Mr Dembo holds field days: NARI brings important visitors to see his farm and these occasions are televised, so that interested farmers can see the quality of his rice fields. And he has paid the radio to announce that he has seed for sale. Some farmers approached NARI to ask for Nerica seed and were referred to him, while Jambur has also referred customers to him once its own stocks were sold.

7.5 Challenges and Strengths of the Seed Enterprises

All three of the enterprises discussed in this chapter are based near to the capital city and the NARI office, and enjoy a close working relationship with NARI. Support from NARI is an important factor in their success, but the benefits do not all flow in one direction. NARI conducts some experiments with the seed enterprises (particularly GHE and its customers) and relies on them for feedback about how the varieties that it promotes work in practice.

GHE benefits from the high level of technical knowledge that its founder contributes, supplemented by its excellent working relationship with NARI. It has diversified into a number of different activities and so enjoys the financial stability that is provided by several distinct revenue streams. Unlike the other two enterprises discussed in this chapter, it began with the aid of a bank loan and continues to use commercial credit when necessary, although most of its financial needs are met with its own funds. This contrasts with the experience of some other seed enterprises, which may thrive without credit but require capital for expansion (as with the Kafo at Jambur). Likewise, Gambian women entrepreneurs engaged in handicraft or horticulture did not need loans for operational capital, but faced problems in accessing capital for expansion (Della-Giusta and Phillips, 2006).

The Jambur Kafo is near the NARI office and has maintained good relations with NARI for a long time. As a result, its members enjoy access to technical advice and new varieties. It is puzzling that despite this link they were unable to renew their foundation seed for many years. The lack of a system to maintain varieties and renew the foundation seed used by seed producers is a systemic weakness across the whole country.

Nevertheless, they intend to use their link with research to enable themselves to respond to changing market conditions. They believe that demand for their main product (seed of 'white Nerica') is declining and so they have developed a realistic long-term business plan that will use the skills that they learned in the PVS, testing and observing different varieties. They are well-placed to obtain new lines from NARI, try them under typical smallholder conditions and sell the best ones. This would allow them to offer a constant stream of new varieties appropriate to farmers' needs. Such a strategy builds upon farmer interest in new varieties and farmer willingness to pay for a promising new variety. But to make the most of this strategy the

Kafo will once again have to use its media savvy to inform consumers about the new varieties it will offer.

One weakness is the Kafo's reliance on the unpaid labour of its membership. In the early years the required labour was forthcoming, since this meant that Nerica seed would become available to each member and so new livelihood opportunities would be opened up for each one. Once supplies of Nerica seed were assured, however, most members gave higher priority to working for themselves rather than for the Kafo, particularly since it is not clear how they benefit from the Kafo's income. Another concern is the Kafo's isolation from agro-dealers. Customers have to come to Jambur, and the Kafo has not created a network to supply larger areas. They thus depend upon motivated customers who are willing to travel long distances to buy seed, and may be missing many other sales opportunities. Nor has the Kafo paid much attention to packaging in order to cash in on their good reputation. If they were to begin to sell through intermediaries, sealed and labelled packets would be essential to protect their brand.

Similarly, Jafaye Farm has not attempted to use packaging to boost sales. Indeed, it chose to ignore the clear demonstration of the power of product differentiation that it was given when some of its seed was sold at a high price after being packaged as a quality product and placed with an agro-dealer. Rather than following this lead, Mr Dembo plans to respond to increased competition by reducing seed production, growing more rice as grain and making the most of his rice mill, which is a scarce and valuable resource.

The business has been successful for several reasons. It began producing Nerica seed early, when demand was high, and for several years could easily sell all that it produced. Its field days have been shown on television and admired by viewers for miles around, while it makes judicious use of radio advertising to bring in business. The prominent location of its main sales point, on a main road, is another advantage that the business enjoys. It operates on a reasonable scale and is adequately capitalized, with several business activities providing the cash flow needed to produce rice seed. The business survived a 50% fall in rice seed production followed by a year when the field was in fallow, demonstrating the resilience that is given by a firm financial base. Mr Dembo owns all the land he cultivates, with enough equipment and good working relations with the providers of technical assistance (NARI and the Department of Agricultural Services) and with supportive fellow seed producers (the Kafo at Jambur).

Both Jafaye Farm and the Jambur Kafo benefit from the activities of companies such as GHE, although they do not know that. GHE brings expertise in marketing and uses techniques such as packaging to present the seed that it resells as being of high quality, while its willingness to sell in small packets from convenient locations makes the seed accessible to a wide range of farmers. As a result, the market available to seed producers is extended, reducing the likelihood that market saturation will in fact become a serious problem for them in the near future.

7.6 Conclusions

The context for the events described in this chapter is the spectacular success of The Gambia in increasing domestic rice production. During 1999–2003, average annual paddy production was just 27,870 tonnes, and yet output reached 38,300 tonnes in

2008 and is projected to increase by 79% to 69,000 tonnes in 2010 (data from the FAO and *Rice Market Monitor*). The impulse that led to this achievement was provided by public policy in the form of a general initiative to make agricultural development a priority, including initiatives personally undertaken by the president, together with the opportunity provided when Nerica became available. The public-service media followed the lead given by the authorities, and the interest shown by their audience prompted them to increase their coverage and also encouraged privately owned mass media to launch related programmes. Their contribution was of immense importance, but it should not be imagined that the media acted autonomously, when they were following an agenda set by the country's political leadership.

The impact of media involvement was in many ways remarkable. The transmission of TV images of fields planted to Nerica made a great impression on several of the farmers interviewed, who were awed by the quality of the plants and immediately decided to grow Nerica themselves so that they too could have fields like the ones shown. The fact that they had seen the fields on TV gave them confidence in the quality of the seeds, and motivated them to travel considerable distances in order to buy seeds from the farms shown on TV. Because of public interest in finding Nerica seed and growing this new rice, it made sense to broadcast more information about why to use high-quality seed and the best ways to grow rice. The media responded, making use of educational materials that were newly available from AfricaRice, and as a result there was a general increase in skill levels and the quality of farming across the rice sector. The overall communication strategy thus evolved from an initial focus on a new variety to broader coverage of the benefits of healthy, high quality seed and good crop management. In Bangladesh, communicating seed health had an equally positive effect on the seed sector (Van Mele *et al.*, 2007).

The mass media fulfilled two important functions: providing information and building trust. This experience confirms the suggestion made by Fafchamps (2004) that strong social institutions can perform a function similar to that of social capital, and so may provide a substitute for it, one that he regards as being preferable on equity grounds.

References

ADF (2005) *Republic of The Gambia Appraisal Report: Farmer Managed Rice Irrigation Project.* African Development Fund, Agricultural and Rural Development Department, Tunis.

Bentley, J.W. (1987) Economic and ecological approaches to land fragmentation: in defense of a much-maligned phenomenon. *Annual Review of Anthropology* 16, 32–63.

Carney, J.A. (1998) Women's land rights in Gambian irrigated rice schemes: constraints and opportunities. *Agriculture and Human Values* 15(4), 325–336.

Della-Giusta, M. and Phillips, C. (2006) Women entrepreneurs in the Gambia: challenges and opportunities. *Journal of International Development* 18(8), 1051–1064.

Fafchamps, M. (2004) *Social Capital and Development.* Discussion Paper Series Number 214. Department of Economics, University of Oxford, Oxford.

Ilboudo, J.-P. (2003) After 50 years: the role and use of rural radio in Africa. In: Girard, B. (ed.) *The One to Watch – Radio, New ICTs and Interactivity.* FAO, Rome, pp. 133–140.

Nuijten, E. (2005) Farmer management of gene flow: the impact of gender and breeding system on genetic diversity and crop improvement in The Gambia. PhD thesis, Wageningen University.

Remington, T.R. and Posner, J.L. (2000) On-farm evaluation of weed control technologies in direct seeded rice in The Gambia. In: Starkey, P. and Simalenga, T. (eds) *Animal Power for Weed Control. A Resource Book of the Animal Traction Network for Eastern and Southern Africa (ATNESA).* CTA, Wageningen, The Netherlands, pp. 255–261. Available at http://www.atnesa.org/weeding/weeding-remington-rice-GM.pdf.

Van Mele, P., Zakaria, A.K.M., Hosne-Ara-Begum, Harun-Ar-Rashid and Magor, N.P. (2007) Videos that strengthen rural women's capability to innovate. *Communication for Development and Social Change* 1(3), 79–99.

Van Mele, P., Wanvoeke, J. and Zossou, E. (2010a) Enhancing rural learning, linkages and institutions: the rice videos in Africa. *Development in Practice* 20(3), 414–421.

Van Mele, P., Wanvoeke, J., Akakpo, C., Dacko, R.M., Maiga Dacko, R., Ceesay, M., Béavogui, L., Soumah, M. and Anyang, R. (2010b) Videos bridging Asia and Africa: overcoming cultural and institutional barriers in technology-mediated rural learning. *Journal of Agricultural Education and Extension* 16(1), 75–87.

8 Morocco: the Visible Hand

Abderrahmane Lyamani, Drissa Silué and Robert G. Guéi

8.1 Introduction

8.1.1 Agriculture

Morocco, on the north-west shoulder of Africa, is bounded by the Mediterranean Sea and the Atlantic Ocean on the north and west, by Algeria on the east and by Mauritania on the south. Its land area of 710,850 km^2 includes agricultural plains and river valleys, plateaux and the Atlas and Rif mountain chains.

In 2007, Morocco's population was estimated at 30.8 million with an annual per capita gross domestic product (GDP) of $2497. Agriculture is the largest sector in the Moroccan economy. Depending on the season, agriculture contributes 13 to 20% of the GDP and over 12% of exports. However, it employs 40% of the working population and provides 80% of rural jobs (Government of Morocco, 2006).

In 1996, there were 1.5 million farms, with an average of 5.8 hectares each. Over half of them (54%) were smaller than 3 hectares, but they made up 12% of the cultivated land and 18% of irrigated land (Government of Morocco, 1999). Scarce irrigation makes smallholder farmers vulnerable to drought.

Of Morocco's 8.7 million hectares of farmland, 7.0 million is rain-fed and 1.7 million irrigated. In an average year, irrigated land contributes 45% of agricultural production, 75% of agricultural exports and 35% of farm employment.

Morocco produces many agricultural commodities, especially cereals, grown on 58% of the cultivated land. Production ranges from 2 to 10 million tonnes per year (an average of 6 million tonnes since 2005). Yet production meets only 62% of domestic demand.

The major limits to agricultural growth are scarce land and water, precarious land tenure, climate change, soil erosion, poor access to credit, lack of farmer organizations, rural illiteracy and the low level of technical training.

8.1.2 Policy

Morocco was one of the few countries that refused to cave in to structural adjustment. It is only now starting to privatize its seed sector.

Moroccan agriculture has received government attention since independence in 1956, with public investment in agricultural development and over 100 dams to irrigate 1.7 million hectares of land. The government also fixes prices for agricultural commodities, inputs and services. The government has provided other infrastructure (roads, highways, railway, airports and seaports), research, training and extension, marketing and credit.

The 2020 strategy for rural development started in 1999. It seeks to increase agricultural production, employment and income, reduce environmental degradation and improve education, social services and infrastructure (MAPM, 2006).

The Green Morocco Plan, developed in 2008, intends to modernize agriculture and integrate it into the world market, improve human development, upgrade natural resource management and enhance sustainable growth. The policy will develop efficient, market-oriented agriculture with private investment for high value commodity chains, and increase the income of the most vulnerable smallholders especially in dry and mountainous zones, through public and private intensification projects allowing a shift to more profitable commodities. The Plan includes reforms of land tenure, water policy and taxation and the reorganization of the Ministry of Agriculture and Marine Fishery (MAPM, 2008a).

The state has subsidized inputs since 1969. In 2004, the government introduced the Agricultural Development Funds (MAPM, 2008b) to encourage private sector investment in agriculture. In 2004 subsidies totalled US$40 million (IFAD, 2008) and covered farm equipment, certified cereal seeds, seed and grain storage and laboratory analyses. This encouragement policy was recently reviewed and new arrangements took effect in January 2010. Two categories of farmers are targeted: small farmers with holdings up to 5 hectares and others with larger holdings. For farm machinery and irrigation systems, more favourable specific advantages are offered to small farmers. As for certified seeds, allocated subsidies are given in Table 8.1.

Table 8.1. Financial incentives to promote seed systems in Morocco. Source: Ministry of Agriculture and Marine Fishery, Morocco.

	Subsidy in US$ or % of total cost
Seed of bread wheat	$19 per 100 kg (40% of sale price)
Seed of durum wheat	$17 per 100 kg (30%)
Barley seed	$17 per 100 kg (40%)
Sugarbeet seed (imported and sold in Morocco)	$875 per unit of 100,000 monogerm seeds
Seed of forage species (maize, sorghum, vetch)	30% of sale price
Sugarcane planting material	$19 per tonne
Seed processing facilities (buildings and equipment)	10%, up to $187,500 per project
Cereal storage facilities (buildings and equipment)	10%, up to $400,000 per project
Cold storage facilities for agricultural products	10%, up to $280,000 per project

8.1.3 Seed systems in Morocco

There are two kinds of seed: certified and common or farmer seed. There is no in-between category. Certified seed is supervised by the Seed and Planting Material Control Service; common seed is unregulated and is distributed with no quality restrictions. However, public subsidies are directed to certified seed only.

Cereal and legume seeds are produced every year on 60,000 hectares, to meet national demand. The use of certified seed in Morocco remains low except for sugarbeet and vegetables (Table 8.2).

Table 8.2. Certified seed use in Morocco. Source: AFSTA, 2006.

Crop	Area sown (1000 ha)	Use (%)
Cereals	4,748	11
Spring cereals	407	10
Legumes	420	3
Fodder	460	10
Oilseed	120	10
Potato	56	30
Sugarbeet	65	100
Vegetables	174	65

The seed system includes national and foreign seed producers, producers' associations, importers and exporters of seed and planting material, research institutions, farmers, government and two national seed traders associations (AMSP – Association Marocaine des Semences et Plants; AMMS – Association Marocaine des Multiplicateurs des Semences). There are a few private seed companies that produce and export seed of some vegetable crops. Although government incentives exist for seed export, the country has no ambition to develop regional seed enterprises.

Seed laws. Morocco is a member of the International Union for the Protection of New Varieties of Plants (UPOV), the International Seed Testing Association (ISTA) and the African Intellectual Property Organization (OAPI).

Morocco has a legal, technical and administrative framework for adequately monitoring the seed and planting material produced or imported in the country, in compliance with international regulations (ISTA, UPOV, OECD and EU). Existing seed laws and regulations cover the protection of new varieties, official catalogue of plant species and varieties cultivated in Morocco, seed control and certification and seed marketing. Major components of the Moroccan seed system are:

Varietal development. The Institut National de la Recherche Agronomique (INRA), a public organization, breeds new varieties, especially cereals, legumes and sugar beets. Varieties of other crops are imported mainly from Europe, USA and Australia. New varieties are submitted for varietal testing and release.

Varietal testing, release and protection. The Seed and Planting Material Control Service decides on release after 2 years of variety trials in at least two locations, compared with local checks. The trials are supervised and approved by a national committee of crop experts. Since 1978, about 5000 varieties have been evaluated and half of them released. The Moroccan official catalogue has more than 1800 varieties adapted to various regions, of which 90% belong to the private sector (AFSTA, 2006).

Seventy crop species are protected in the country and 240 varieties originate from INRA, of which 219 are from annual crops (Table 8.3). The plant breeding companies pay 500–1000 dirham ($60–$120) for each registration in the catalogue.

Table 8.3. Achievements of INRA annual crops breeding programme. Source: INRA

	Number of released varieties
Durum wheat	35
Bread wheat	25
Barley	24
Oat	17
Maize	21
Rice	17
Broad bean	6
Chickpea	11
Lucerne	3
Vetch	9
Lentil	9
Bean	6
Cotton	9
Sunflower	4
Others	23
Total	219

Seed multiplication. INRA produces breeder seed for its own varieties. Private or national seed growers produce foundation seed and certified R1 and R2 seed of these varieties. One parastatal organization, the National Seed Marketing Company (SONACOS – Société Nationale de Commercialisation de Semences, which became privatized in 2007), multiplies seed of most INRA varieties and pays royalties based on commercial seed sales, up to 2.5% for wheat and 1.5–2.5% for food legumes. SONACOS also pays a lump sum for every newly released variety.

Seed processing. This includes drying, cleaning, sorting, calibrating, treating, packing and labelling. INRA has five units to process breeder seed. SONACOS and other seed enterprises have facilities to process foundation and certified seed.

Seed quality enforcement and certification. The Seed and Plant Material Control Service (SCSP – Service de Contrôle de Semences et Plants) is the official agency in charge of seed certification at all stages of production, processing, storage and marketing of seed. Seed is inspected in the field, laboratory and postharvest. The Service has 129 employees and 14 regional offices.

Monitoring ensures that new varieties remain high yielding, adapted and resistant to disease. It also protects breeders' rights to encourage variety owners to provide the Moroccan market with their best varieties, providing farmers with good, healthy seed and planting material.

Field inspection is the hallmark of certified seed.

Specific seed regulations cover cereals (durum and soft wheat, barley, triticale, oats, rye and rice), food legumes (broad beans, peas, lentils, chickpeas and beans), forage legumes (lucerne, clover, purple clover, forage peas, vetch and lupin), sugarbeet and forage beet, cotton, maize, oil crops (sunflower, safflower, rapeseed, flax, soybean and groundnut) and standard seed (mainly vegetables).

Seed certification is based on both field and laboratory results. Ten thousand samples, all crops combined, are analysed per year. Farmers pay 2.5 dirham ($0.30) for each field inspection and the government covers the remaining cost. Once a seed lot is certified, all bags are sealed with tags inside and outside.

Market promotion and distribution. To obtain a marketing agreement one must have a professional qualification in seed or employ a qualified technician, deal in certified seed of varieties registered in the official catalogue and have a suitable storage facility.

Certified seed is distributed and sold through a network of more than 92 accredited establishments (Table 8.4). Most of them are members of the AMSP seed traders association, created in 1991.

Private seed companies import and export seed of food and forage legumes, maize, oilseed crops and vegetables. Seed of these crops is mainly used in intensive irrigated production systems. Some European companies produce seed of beans, peas and vegetables (onion and carrot) in Morocco and take their production home for processing and marketing. The parastatal seed enterprise, SONACOS, has roughly the monopoly on wheat, barley, rice, potato and sugarbeet. For sugarbeet, all certified seed is imported from Europe. Morocco tried to produce sugarbeet seed in the mountains but the cost was so high that this option was discarded.

Cereal seed prices are set annually at the beginning of the growing season. Seed producer prices are based on the costs of production, transportation and storage plus a profit margin. Sale prices are also set with an assumed fixed profit margin. In 2009, certified cereal seed received a subsidy of 30% of sale price.

Table 8.4. Distribution of private seed companies in Morocco.

Location headquarters	Number of seed companies	Crops for which certified seed is produced	Number of seed companies
Casablanca	54	Food legumes	68
Rabat	11	Maize	60
Fès	8	Oilseed crops	50
Kenitra	7	Forage legumes	44
Agadir	3	Vegetables	35
El Jadida	3	Wheat and barley	9
Sidi Bennour	2	Potato	4
Berkane	1	Rice	3
Meknes	1	Sugarbeet	1
Nador	1	Forage beet	1
Beni Mellal	1	Cotton	1
Total	92		

The National Seed Plan adopted in 1997 identified the main limitations for the use of certified seed: climate, drought, price, low profitability and inadequate varieties, extension and organization. Climate change and recurrent droughts lead farmers to manage risk by reducing their investment in agriculture. Farmers' cash supplies evaporate after several years of drought. For major staple crops (cereals and legumes), after a good harvest, farmers tend to use their own seed. Also, they buy less certified seed when the price exceeds 1.5 times that of farm-saved seed. Seed enterprises have low profitability, especially for seed of open-pollinated crops. There is a shortage of varieties that meet the requirements of some specific agro-ecological zones such as mountains and irrigated land, and insufficient extension and promotion of certified seed and improved varieties. The inter-professional seed organizations are poorly developed.

The National Seed Plan aims to revitalize and develop the seed sector and set up an inter-professional seed organization able to coordinate the seed industry and ensure a steady market supply of certified seed.

8.1.4 Evolution of the seed industry in Morocco

The seed industry has gone through three major phases:

From Independence in 1956 to the late 1970s. The seed industry was composed of state farms and experimental stations, and private enterprises for producing, processing and marketing seed, especially for wheat and food legumes. The government created SONACOS in 1975 to secure seed production of staple food crops. SONACOS developed a network of seed multiplication farms including SOGETA (Société de Gestion des Terres Agricoles), a state company which manages state farms recovered after independence.

Early 1980s to late 1990s. Seed laws were enacted and breeding programmes started. Private seed companies began showing more interest in importing horticultural seeds.

Early 2000s to the present. SONACOS created regional centres with modern seed-processing and storage facilities. New private seed companies including international ones showed interest in Morocco and some of them started private breeding and seed production for cereals, maize, tomato, potato and other crops. The government created the Agricultural Development Funds to encourage private investment in agriculture and to subsidize certified cereal seed. In 2007 the government made available 11,754 ha of state-owned land for five seed projects as a public–private partnership. This land was allocated to Moroccan and foreign seed enterprises with good track records of producing certified seed.

8.2 The SONACOS Seed Enterprise

8.2.1 History

SONACOS was created in 1975 and was 100% owned by the government. The headquarters are in Rabat with regional centres nationwide. To ensure sales everywhere, ten regional wholesale centres and 380 retailer shops were established across the country.

The government started SONACOS to handle seed for bread wheat, durum wheat and barley, which are the main food crops. Later the enterprise began producing seed of other crops, including potato, sugarbeet, fodders, food legumes and oil crops. SONACOS currently controls 95% of the seed market, including all of the sugarbeet seed.

For a token payment, SONACOS received rights to all the varieties developed by INRA, including mainly cereals and food legumes. After the mid-1980s, all INRA varieties were tendered on bid, but the royalties were still low. SONACOS is allowed to introduce varieties from outside Morocco if they are high yielding or have specific traits, such as disease resistance and quality.

8.2.2 Structure

Management. The company has a board which controls all activities, including finance and management. It has five departments: commercial, technical, financial and accounting, equipment and maintenance, and auditing and management control. SONACOS also has an administrative service and a central purchasing unit.

In 2007 the government instructed SONACOS to operate as a private enterprise, meaning that the company should recover all costs, including salaries. The company shed many jobs but still employs more than 220 employees.

Land and labour. SONACOS owns no land, but subcontracts outgrowers, who produce seed with machinery (e.g. tractors and seed drills) and agrochemicals.

Outgrowers are free to use any technology, as long as they respect the technical regulations for seed certification.

Quality control. Outgrowers buy their own foundation seed and produce certified seeds, with quality control by official SCSP inspectors. SONACOS only buys the final certified seed and the price is based on quality. Outgrowers know the seed price since planting and can sell to SONACOS once their seed has been certified.

Infrastructure and equipment. SONACOS has a huge storage capacity of 100,000 tonnes for cereals, a seed treatment and packing capacity of 120,000 tonnes, a cold storage facility at Larache of 5000 tonnes for seed potato, and seed treatment and packing equipment. The company also has two stations for evaluating and demonstrating new varieties.

Access to high performing varieties. SONACOS sells certified autumn and spring cereals, fodder legumes and grasses, food legumes, sugarbeets, potatoes, oilseed crops and vegetable seeds. SONACOS signed agreements with national and international plant breeding organizations to gain access to high performing plant material. SONACOS works closely with INRA and obtains varieties by paying a symbolic royalty that covers only some of the costs of breeding.

Trials of varieties registered in the national variety catalogue are conducted each year in different agro-ecological zones. These varieties are promoted at open houses, agricultural shows and field days.

Training. SONACOS trains its staff and personnel of public and private organizations on seed multiplication. It trains outgrowers on seed regulations and policies, sale of seed and farm inputs and quality control. In 2008 SONACOS began dealing in agrochemicals and selling its expertise to other organizations.

Farmers help select new varieties of broad bean, identifying the best ones for their conditions.

8.2.3 Cash flow

In 2009 SONACOS had a capital of 3 million euros ($4 million) and a turnover of 50 million euros ($70 million).

SONACOS has access to government-guaranteed loans and is now modernizing its storage facilities. With 95% control of the seed market, the company has little fear of competition, because private companies are uninterested in the crops covered by SONACOS.

The government is now considering opening the company shares to private investment and believes that this will improve the image of SONACOS and create fair competition in the seed sector.

8.2.4 Marketing

SONACOS supplies quality seed nationwide to farmers, projects, organizations, NGOs and government agencies. In 2009, it also provided expertise in seed marketing in Senegal.

Customers buy seed from SONACOS because of its nationwide presence and good seed of adapted, high-yielding varieties. Seed is advertised on radio and TV. Cooperatives and entrepreneurs are using similar marketing strategies in Mali (Chapter 5) and The Gambia (Chapter 7).

SONACOS is thinking of establishing value chains for some crops, starting with cereals, and including wheat growers, the seed industry, millers, flour distributors and retailers. The company believes that its most serious competition is from informal seed and not from emerging private seed companies.

8.3 Challenges and Strengths

When the International Monetary Fund (IMF) forced most other African countries to toss their seed industries into the invisible hand of the market, Morocco resisted. Thanks to government support, SONACOS has had well-functioning seed production, conditioning and marketing systems. Its 14 well-equipped regional seed centres place Morocco among Africa's leading nations in the seed industry. Thirty-four years' experience in seed is another advantage. Seed production and use is subsidized up to one-third of the seed price. Farmers use certified seed and inputs because of their low cost. If they had to pay a higher price, farming might not be profitable. But the government of Morocco is now encouraging SONACOS to recover more of its costs and become more of a parastatal corporation than a government agency. There are other African countries with long-standing parastatal seed companies, including Ethiopia, Kenya and even the ADPs of Nigeria (Section 4.3), still guided by the visible hand of public servants.

References

AFSTA (2006) News from the Seed Trade Association of Morocco (AMSP) Electronic newsletter of the African Seed Trade Association, June 2006. The African Seed Trade Association, Nairobi, Kenya.

Government of Morocco (1999) *Moroccan General Census of Agriculture, 1996.* Rabat, Morocco.

Government of Morocco (2006) *Cinquante ans de développement humain et perspectives 2025.* 50th anniversary report 2006. Rabat, Morocco.

IFAD (2008) *Kingdom of Morocco, Country Strategic Opportunities Programme.* Document EB 2008/95/R.15, 28 November 2008.

MAPM (2006) *Implementation Status of the 2020 Strategy for Rural Development.* Ministry of Agriculture and Marine Fisheries/World Bank/FAO, June 2006. Rabat, Morocco.

MAPM (2008a) *Green Morocco Plan: Integrated Development Strategy for Agriculture in Morocco.* Ministry of Agriculture and Marine Fisheries, Rabat, Morocco.

MAPM (2008b) *Les aides financières pour l'encouragement des investissements agricoles privés.* Ministère de l'Agriculture et de la Pêche Maritime, Rabat, Morocco.

9 Kenya: a Company, a Cooperative and a Family

MICHAEL MISIKO, CONNY ALMEKINDERS, IAN BARKER, DINA BORUS, JUDITH OGGEMA AND JOHN MUKALAMA

9.1 Introduction

9.1.1 Agriculture

Eighty per cent of the people of Kenya make a living from agriculture, but for 60% it is a precarious life of survival on less than a dollar a day. New, appropriate technology would increase agriculture's current contribution of 24% to the gross domestic product (Republic of Kenya, 2009).

Kenya can be divided into four agro-ecological zones, namely: (i) high potential (a) mixed farming (b) cereal and dairy; (ii) marginal agriculture; (iii) agro-pastoral; and (iv) mostly pastoral (Republic of Kenya, 2009). The enterprises described in this chapter are all in the high potential, high altitude zone, with relatively high rainfall, high population density and constant migration. Here most seed technologies have been tested and adopted since the colonial period (Misiko, 2007).

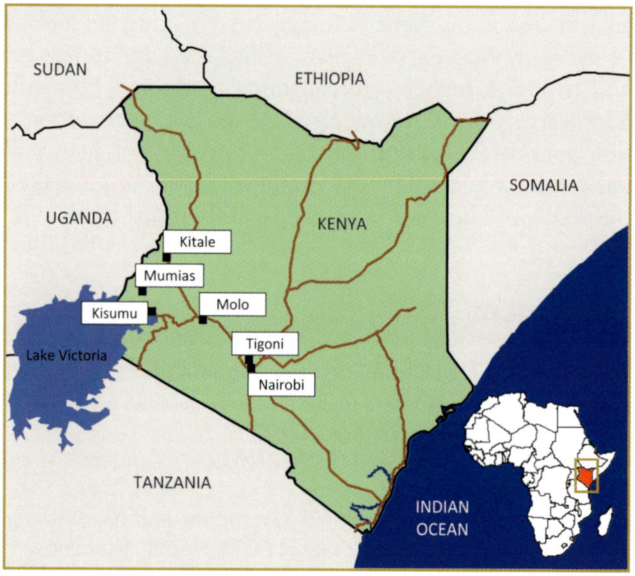

Since the colonial period, maize and beans have been popularized in Kenya along with cash crops such as coffee and tea. After independence, policies have neither evolved nor been implemented to promote other crops aggressively (Misiko, 2007). Compared with maize, groundnut or rice with established seed and commodity marketing channels, soybean, for instance, has a long way to go in improving the marketing links between farmers in rural areas and processers, both large and small (Misiko, 2007; Tinsley, 2009).

On the other hand, potato has enjoyed a vibrant existence especially because of its demand in both rural and urban centres. This vibrancy, however, is often curtailed due to the unavailability of quality seed.

Healthy seed of high yielding varieties will be key to increasing agricultural productivity. While the world seed market is worth about $30 billion, sub-Saharan Africa's share is a mere $800 million, 3% of the world's total. Kenya is a middle range key player with a $42 million seed market (Republic of Kenya, 2009), next to South Africa ($160 million), Morocco ($160 million), Egypt ($140 million), Nigeria ($120 million), Zambia ($15 million) and Malawi ($10 million).

9.1.2 Seed systems in Kenya

Informal sector. The diverse informal seed sector obtains planting material from farm-saved seed, farmer-to-farmer exchange, local markets, NGOs, community-based organizations (CBOs), entrepreneurs and the formal sector. The informal sector includes any seed producer, company or dealer who is not registered with the Kenya Plant Health Inspectorate Service (KEPHIS), including relief agencies and major flower firms who supply seed of unknown quality.

Formal sector. The formal sector includes donor agencies, KEPHIS, the Plant Breeders Association of Kenya (PBAK), the Seed Trade Association of Kenya (STAK), agents, sub-agents and agro-dealers. It is broadly governed by several laws, namely the Plant Protection Act (CAP. 324); the Agricultural Produce Export Act (Cap. 319), the Seeds and Plant Varieties Act (Cap. 326); and the Seeds and Plant Variety (NPT) Regulations, 2009. These laws are enforced by different agencies, besides KEPHIS (Republic of Kenya, 2009).

KEPHIS is the regulatory body established in 1996. It has played a critical role in streamlining operations of the seed industry to ensure that farmers have access to certified seed. It has overseen the growth of seed companies operating in Kenya, from a mere 18 in 1996 to 73 in 2010. This growth has seen local seed companies increase and dominate local seed business. Over this period, certified seed production has tremendously improved. For instance, the rejections in seed crop fields during certification reduced from 30% to 15%, while seed lots not meeting national standards have recently reduced from 15% to 4% (www.kephis.org).

Most of the Kenyan seed companies sell seeds for cereals, oil crops, pulses, pastures, and fruits and vegetables, mostly crops which also dominate research in the Ministry of Agriculture, KARI, the African Insect Science for Food and Health (ICIPE), International Potato Center (CIP), International Center for Tropical Agriculture (CIAT) and other research institutions.

Producers of open-pollinated varieties are CBOs, small or medium-sized enterprises or small units at the Kenya Agricultural Research Institute (KARI). Large companies that have ventured into OPVs have either opted out or scaled back due to low profits. CBOs and national institutions such as KARI and the Agricultural Development Corporation (ADC) are driven more by social responsibility than by profit, i.e. to grow quality seed for family farms. The bulk of seed certified by KEPHIS is hybrid maize (Table 9.1).

Table 9.1. Seed of crops certified (tonnes) in Kenya, 2004–2009.

	2004	2005	2006	2007	2008	2009
Maize OPV	562	464	244	473	2	138
Maize hybrid	23,686	22,545	26,987	27,989	22,698	29,264
Beans	153	164	179	307	457	378
Sorghum	506	350	539	544	606	3,215
Millet	50	65	33	8	13	27
Soybean	0.1	–	0.5	2	–	–
Groundnuts	–	1	11	2	–	4
Cowpeas	191	89	103	176	145	166
Green gram	–	12	25	36	115	124
Pigeon peas	21	19	7	8	4	8
Potato	93	62	410	225	362	489
Total	25,322	23,922	28,719	30,159	24,606	33,912

9.1.3 Policy

The seed sector in Kenya is regulated through the Seeds and Plant Varieties Act, which allows any producer, processer or marketer of seeds that meets the standards and requirements to register as a seed enterprise (KEPHIS, 2008).

Seed certification. KEPHIS certifies seeds of both hybrid and open-pollinated varieties. Most material submissions to KEPHIS come from KARI (57 submissions), the Kenya Seed Company (41), Leldet Ltd (17), the University of Nairobi (15) and Western Seed Company (six). KEPHIS handles mostly maize seed, whereas potato accounts for less than 3% of the entries.

Potato is logistically complex to inspect and certify because the fields are often small and scattered and the bulky seed samples are difficult to carry back to the lab. Sampling and analysing are important because potato has many seed-borne diseases, so seed certifiers often have the unpleasant task of rejecting a smallholder's seed plot because of disease. The main institutions producing and distributing certified seed potato are the ADC and KARI-Tigoni, while soybean seed is handled by CIAT.

Regional harmonization. Kenya belongs to several African associations that support agriculture, including the Association for Strengthening Agricultural Research in Eastern and Central Africa (ASARECA), which has as one of its projects the harmonization of seed policies and regulations among member countries. However, Kenya claims to be the only Eastern African country that follows OECD seed standards. Kenya's seed testing laboratory is accredited according to the strict rules of the

ISTA – International Seed Testing Association (KEPHIS, 2009). The lack of compliance among the other countries hampers formal seed trade in the region.

9.1.4 Seed use and demand

Maize. Sixty per cent of Kenyan farmers use certified maize seed, mostly hybrid, at least some of the time. Seed enterprises see OPV maize seed as unprofitable and few companies produce it; 80% of the total supply of maize seed comes from the informal sector. Most Kenyan farmers usually keep their own seed, often selected from hybrid harvests, to be used in two or even more subsequent seasons (Mango and Hebinck, 2004).

In 2008, only 9% of Kenya's seed imports were maize, mostly varieties not found locally, but 76% of all seed exports were maize as well (KEPHIS, 2008) because Kenyan maize seed was competitive in the EU, Japan and elsewhere.

Soybean. The national demand for soybean or derived products is at least 100,000 tonnes per year (Tinsley, 2009), mostly for industry (e.g. for animal feeds) and for schools and refugee camps. Food in which soybean is an ingredient is not in demand by households. Most soybean in Kenya is imported from Uganda, Malawi, Brazil and Argentina, so capacity building along the value chain is needed (Misiko, 2007).

Potato. About 59% of the farmers never renew their seed potatoes. Those farmers who do renew their seed potatoes do so after an average six seasons, so only 7% of the seed stock of Kenya gets renewed each season from any of the possible sources outside the farmer's own farm (Gildemacher *et al.*, 2009). Seed potato exports were modest, even though potato is the second biggest crop in Kenya, because only about 1% of seed potato required was certified, far below domestic demand (Guyton *et al.*, 1994; Republic of Kenya, 2007).

Case studies. This chapter presents three case studies. The Western Seed Company is one of the few seed companies in Kenya that produces OPV maize, besides hybrids. Western Seed produces seed of several varieties of OPV maize, showing that it is possible to run a large, successful OPV seed business, even if it is a challenge. Second, the Mumias District Federation of Soybean Farmers (MUDIFESOF) is a promising CBO, the only one expanding its seed production. It started as a CBO before producing seed and enjoys a lot of support from a research institution, both of which are probably important for its success. And, third, Sungus Enterprise is a thriving family business that produces clean seed potato.

9.2 Western Seed Company

9.2.1 History

Western Seed Company Ltd was founded in 1986. In 1990 it started producing seed of OPV maize varieties grown in western Kenya as well as new varieties from national research. It targeted smallholders by producing seed of robust, early maturing, disease-tolerant varieties, adapted to dry areas and wide agro-ecological zones.

Western Seed now reaches 200,000 farmers through its network of distributors. All of its seed is produced in Kenya, including hybrids. Most of Western's varieties are developed through its own research programmes or through collaboration with research institutions such as ICIPE and KARI.

Farmers prefer the company's hybrid maize WH502 over other varieties because it is resistant to the parasitic striga weed, maize streak virus and grey leaf spot and is drought tolerant. Western Seed has many other successful hybrid maize varieties, better known than any of its OPVs. However, Western Seed has produced several of its own OPV varieties, besides researching and producing sorghum, cowpea, common bean, pigeon pea, sunflower, amaranth, green gram and millet. Besides its own lines, Western Seed is successfully collaborating with KARI and other research institutions to market other OPVs (www.westernseedcompany.com).

According to its director, Syed Osman Bokhari, the OPV varieties earn the company less than 5% of its profits. In 2009, 400 tonnes of WS909 OPV maize seed were earmarked for an NGO, which failed to collect it. The seed was burned to avoid extra costs of storage (which needs extra electricity and manpower).

Mr Osman explained that farmers' preferences for hybrid seed were based on logical calculations of benefits. For instance, 2 kg of OPV seed costs KSh 200 ($2.70) and hybrid seed costs KSh 240 ($3.20). When planted on 1 acre (4000 square metres), a farmer needs 10 kg of seed and will therefore only spend an extra KSh 200 ($2.70) on seed when planting a hybrid. This, however, translates into an extra average of 15 bags (1350 kg) of maize yield per acre, representing an average KSh 35,000 ($467) more in the value of the harvest per acre, at recommended rates of fertilizer. The yield differences between certified OPVs and farmer-saved seed under smallholder conditions are not documented, though farmers in Mumias believe there is little difference.

Western Seed Company has been in business for about 25 years. Although it started producing OPV maize, this now counts for less than 5% of its profit.

Though they are not as profitable as hybrids, Western Seed discovered that OPVs have a unique niche when adapted to smallholder fields plagued with drought, infertile soil and the striga weed. Western Seed explains that it is comparatively easier and even cheaper to deal in OPVs with existing extra advantages, rather than to develop hybrids with similar or better capacities. The company is therefore investing in research to develop quality protein OPV maize varieties (e.g. WS104), adding qualities that do not already exist in OPVs. These OPV lines are developed as part of the company's corporate social responsibility, reflected in their slogan of 'bringing technology to the farmer'. OPVs turn out to be a successful marketing scheme (they are cheaper and offer added advantages), now enjoying brisker sales among smallholders, who are 99% of the company's clients.

9.2.2 Structure

Management. Western Seed Company has 50 staff, a managing director and two directors who oversee five departments: marketing, research and farming, processing, administration and sales support. There are no separate departments for different crops or for OPV and hybrid varieties. They have many agents and agro-dealers around Kenya.

Outgrowers. KEPHIS inspects and certifies seed of the company's outgrowers, a few hundred large-scale farmers and companies in Western Kenya near their facilities, who have proved that they can honour contractual obligations. Western Seed Company provides foundation seed, training and follow-up and buys back seed at competitive prices. The outgrowers are not organized as a group.

Land. The company leases thousands of hectares of land for research on their own and with research centres and for growing seed. Most of the seed land is in Trans Nzoia district, which has fairly good roads leading to Kitale town, where Western Seed is based.

Infrastructure. Western Seed has all the necessary facilities, including modern laboratories and seed processing units, huge storage facilities, ample machinery and a packaging unit. The company maintains multi-tonnage lorries and other vehicles, specialized field production equipment and loading facilities. From its headquarters in the small, agricultural city of Kitale, Western Seed has access to roads, electricity, suppliers and research. Western Seed invests constantly in upgrading its facilities.

Maintaining and upgrading the infrastructure and producing the seed that farmers want are key to staying in business.

Farmers' feedback. The company gets smallholder feedback through its extensive network of seed dealers and by interacting with farmers during field days, demonstrations and open discussions. This allows them to learn about farmers' needs for new varieties and other corporate services (see Box 9.1).

9.2.3 Cash flow

Western Seed Company faces competition from multinationals, CBOs, the Kenya Seed Company and farmers. It adds or drops seed of particular varieties, usually based on research and sales statistics. It keeps strong capital reserves to manage risk. It does not have insurance for its seed farms or for outgrowers. All seed ventures are prone to floods, drought and theft, driving up insurance premiums.

Western Seed estimates that seed demand in Kenya is increasing and still unmet. This is based on three observations: first, their distributors finish their stocks early and clients are still looking for seed; second, the company's output has been increasing steadily; and, third, the surveys done by their own marketing department. So even with the existing competition there is little worry for a well-run company.

Western Seed Company's main concerns are production costs (electricity) and targeting the right clients by producing the seed they want rather than simply producing good seed. It has strong flagship varieties so they can target certain niches. For instance, WH502 is preferred for its high yield. Farmers in Mumias said it 'is voluminous' and good for the market. Most farmers in Kenya sell their produce in 2 kg tin or plastic containers (locally called *koro-goro*). So farmers sell maize by volume, not by weight. Farmers also like WS502 because it is adapted to local conditions, e.g. low rainfall and the parasitic striga weed. Its main competitor is the hybrid 614D from the Kenya Seed Company, which is popular because of its taste and heaviness – it satisfies hunger. Western Seed targets client niches (e.g. for market, for roasting) for farmers who do not necessarily seek more seed but want seed with unique advantages.

Box 9.1 The ideal variety: does it exist?

On 2 February 2010 Michael Misiko met with a group of maize farmers in Mumias to learn why open-pollinated varieties (OPV) from Western Seed Company are used by farmers. Farmers easily recalled the certified OPV Katumani. Informants remembered the hybrids more, especially the striga-resistant WH403 and WH502, which they preferred for its high yield, big cobs, uniform bright colour and drought tolerance. But WH403 was not available.

Nor did the farmers know well how to tell Western Seed that WH502 rotted after heavy rains. 'It opens at the tip ("mouth") and is not good in long rains.'

The farmers also liked hybrid 614D from the Kenya Seed Company, because it is 'heavy, and good tasting'. Participants were willing to pay extra for it, and for WH403. They grow well during the first season, and at higher altitudes.

The farmers appreciated all three of these varieties and did not rely on any one seed source or type; none of them was the ideal variety.

They had all learned about the varieties at demonstrations, from FM stations and by word of mouth.

Western Seed Company believes that seed has to be affordable and being near its main buyers in western Kenya is a huge advantage. Most of the smallholders who buy its seed are in North Rift, Western and Nyanza Provinces. Western Seed also prices its seed to reflect its quality and its advantages, such as resistance to striga and high yielding capacity.

Cash sales only. Western Seed Company is the only seed company in Kenya that prints recommended prices on its packages. The company's annual customer base of 200,000 farmers is growing. The sales department ensures that the seed prices do not discourage buyers while still ensuring a profit margin for agro-dealers. Wholesale prices vary depending on the reliability of the agro-dealers and their annual seed turnover. Sales representatives earn a commission on each seed sale and Western Seed staff monitors seed circulation through visits. Retail prices are the same all over Kenya, since smallholders insist on paying the price on the seed package. Western Seed does not give agro-dealers credit; all sales are cash and carry. This is quite an achievement especially compared with Nigeria (Chapter 4) where companies are forced to sell most of their seed on credit.

9.2.4 Marketing

Western Seed Company does not retail any seed outside its facilities in Kitale. Marketing was simple when the company was young. As the company grows, it relies more on a network of distributors and agro-dealers, who are all trained by the company. It also relies on its reputation and trust from clients. Therefore the main task of the marketing team is to ensure that the company brand is unique and known to as many people as possible, through the media and marketing channels, but more critically through effective communication with the government, KEPHIS, research and development partners and especially farmers. The company is also helped by client loyalty and by word of mouth advertising.

On-farm demonstrations provide technical advice for smallholders on seed, fertilizer, agrochemicals, etc. Field training strengthens loyalty and retains old customers. Other information channels are Agricultural Society of Kenya (ASK) shows, open community (*barazas*) meetings, farmer field schools, leaflets, field days, open days, newspaper and radio – especially regional vernacular ones such as West FM.

Western Seed's reputation helps to spread every new variety that is released; clients just want to try it. The most successful agro-dealers are usually the ones who help push new releases. Recognizing this, the company's staff trains them to use best business practices and to provide useful feedback for Western Seed.

Western Seed Company's consistent and recognizable packaging allows farmers to identify their seed, reducing scope for counterfeiters and cutting packaging costs (e.g. by minimizing redesign costs). The company has kept a specific label and colours for a long time, and uses a particular type of thread and paper material. Customers and government agents easily recognize the simple package, which is duly certified and stamped by KEPHIS (KEPHIS, 2009: 6). The company adheres to regulations, simplicity and client loyalty to beat counterfeit seed.

9.3 Mumias District Federation of Soybean Producers

9.3.1 History

Mumias District Federation of Soybean Producers (MUDIFESOF) is a young, community-based organization (CBO) with 31 members. It became known as MUDIFESOF after researchers identified it and when it was registered under the Social Services Department of the Kenya government in 2007. Before this, its members collaborated in many other activities, including seed production. The Federation works closely with several supporting agencies, especially CIAT, but also the Ministry of Agriculture, KARI and the Western Region Christian Community Services. This case illustrates the critical role of community-based seed enterprises, and why they may run out of steam over time.

MUDIFESOF is based on organized leadership and on the technical and financial support of CIAT for soybean varieties, in the form of training and $3000 annual support (ended in 2009). CIAT supports the introduction of new soybean lines, especially high yielding ones. MUDIFESOF supplies its seed to thousands of farmers from Mumias and beyond, including those working with KARI, the Lake Basin Authority and other CBOs. Soybean is a relatively new crop among smallholders, introduced into Kenya in the 1970s, but never popularized.

9.3.2 Structure

Management. MUDIFESOF's 31 members include a chairman, a secretary, a full-time caretaker (who looks after the buildings and equipment) and five representatives in Mumias District (one in each division), who mobilize seed growers, some of whom are not members of MUDIFESOF. CIAT gave all the members training in seed production, while KARI taught them processing and marketing. The members have taken further training as needed, e.g. on new diseases, pests or agrochemicals. MUDIFESOF outgrowers (non-member seed producers) have not all been trained and have to be closely supervised while the crop is in the field. This involves roguing off-type plants.

In 2009, MUDIFESOF produced 2 tonnes of soybean seed which was uncertified, because they are not registered with KEPHIS. The five representatives monitored farmers' fields to see that seeds are not mixed. MUDIFESOF provides quality seed of new lines, based on customer variety preferences, determined by highly valued variety characteristics (e.g. amount and taste of soya milk) and maturity rate. Varieties that can be stored for up to 12 months are preferred over ones that last only about 6 months.

Infrastructure and equipment. MUDIFESOF's storage facility is a simple room, with no temperature regulation and with no special storage bags (e.g. sisal). Bags usually carry handwritten labels that indicate the variety names. Seed characteristics are not further specified, but are told to farmers along with other verbal explanations on crop management practices.

MUDIFESOF relies on mobile telephones, and it is connected to the national electricity grid. The limited storage capacity is not a serious obstacle as the seed moves fast, especially because buyers do not line up at planting only but rather buy seed early to store it themselves. The most sought-after varieties are still under-supplied and are usually bought immediately after harvest.

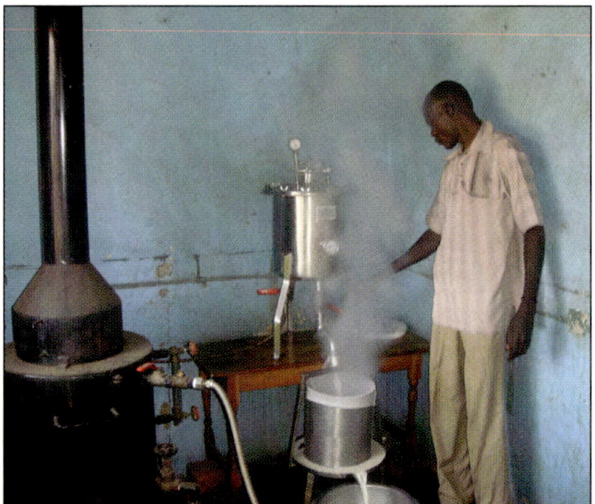

The community group in Mumias has received ample support from research to improve the use of soybean by rural communities. Apart from processing, the group has also embarked on seed production.

The CBO has basic sealing equipment and a standard kitchen fridge for storing about 30 litres of soya milk. It has a machine to make soya milk and other soybean products. MUDIFESOF's demand for grain from farmers is increasing. This grain is processed into products such as milk and flour. This in turn has led to the growth of demand for soybean seed, which is supplied by MUDIFESOF. A similar positive value chain effect on the seed market was observed in Nigeria, where a brewery needed a regular supply of sorghum of a particular variety (Section 4.6).

9.3.3 Cash flow

MUDIFESOF sells most of its seed for cash, but sells a small amount to trusted farmers on credit. They also sell soya milk and soya flour to local consumers for cash. They receive soybean grain from farmers. If they have sold seed on credit, farmers can repay in grain, although MUDIFESOF prefers cash sales, which are safer. MUDIFESOF also processes and packages other grain for farmers, and can label these soybeans in the other farmers' names. This provides cash for MUDIFESOF, but gives potential competitors a leg up. From 2007 to 2009 MUDIEFESOF also received $3000 per year from CIAT, and still receives some support from KARI, for training, and this helps them sell seed to other CBOs.

9.3.4 Marketing

MUDIFESOF more than doubled its seed sales since 2007, mostly to smallholders. Their selling point is straightforward; the new varieties can fix nitrogen without inoculation, are tasty and high yielding, suppress striga weed and generate income. Besides, the varieties are from renowned research institutions. KAPP and MoA had several field days, promoting soybean as nutritious food for HIV patients and children.

MUDIFESOF has a front office for sales, promotion and collaboration. Other CBOs are copying MUDIFESOF's style, but they are still not serious competitors and demand for soybean seed is growing. This demand is reflected in the rapid growth of MUDIFESOF seed production: 2007 – 1.5 tonnes, 2008 – 2.7 tonnes, 2009 – 5.4 tonnes and in 2010 it expects to have 7.5 tonnes. Although this seed is not certified, a growing number of clients were associating MUDIFESOF with certain varieties, especially SB25 (Namsoy) and SB3, demand for which has soared since 2007.

The price of seed should be at least twice that of grain, considering the inputs required and the benefits of quality seed. In February 2010 a kg of soybean seed at Mumias town was selling for KSh 80 ($1.10), while grain was half that price. Because soybeans are new in Kenya, few enterprises deal in them.

MUDIFESOF built its reputation by giving seed on credit to smallholders, to be repaid in cash or in soybean grains. But customers may mix seeds, a serious challenge beyond merely repaying the debt. This CBO is so dependent on the community's goodwill that they never reject seed from a producer. Mixed or low quality grain is often processed and sold as another product, not as seed. MUDIFESOF also offers packaging and processing services for other community businesses like that of a local soyabean entrepreneur named Alice. The risk is that MUDIFESOF is helping Alice and the others become more serious competitors.

9.4 Sungus Seed Potato Enterprise

9.4.1 History

Sungus Enterprise produces clean seed potato in Molo District under the management of Anne Mbugua, the wife of Mr Sungus. Seed potato is part of a larger family business that also includes ware potatoes, a dairy farm, a registered hotel and other activities. But as recently as 1985 the Sungus family just had a conventional potato farm. It steadily evolved until 2005, when Anne drastically transformed it after taking a GTZ-led 'clean seed potato production' course. Mrs Mbugua qualified for the GTZ training because of her farm's potential to produce clean seed, e.g. she uses about 2 hectares of land at a time to allow for crop rotation.

9.4.2 Structure

Management. Sungus is a family enterprise, with 15 regular employees producing seed potatoes. More workers are hired for harvesting, mostly people who have worked for Sungus before. The permanent and temporary workers receive all their training on the job. Only one variety may be planted per plot at a time, which happens monthly. After two potato crops Sungus plants oats and feeds them to the dairy cows. This helps to cut down on costs and disease and reduce volunteer potato plants from the field, so that Sungus can plant a different potato variety after harvesting the oats.

Sungus values hard work and commitment. Anne has learned to use positive selection, which means identifying the best plants in the field and keeping their tubers for seed for the next season. The healthy plants are marked with a peg and harvested separately. She also uses negative selection, removing sick plants. Such selection requires patience and skills.

Sungus is a family enterprise devoted to producing quality seed potato.

In recent years, Sungus has mainly produced Sangi seed, i.e. seed of a newer variety whose origin and source is not clear. Sungus bought Sangi from a seed seller and multiplied it. Some people perceive Sangi as unscreened and dangerous, of poor storage abilities and from an unknown source (Organic Farmer, 2009). But farmers like this red-skinned variety because they say it is tolerant to poor soils, high yielding, disease resistant, early maturing and tasty. It is now Sungus's best-selling variety. But Anne thinks demand for Sangi may decline. It sprouts fast, so it has to be sold quickly, or distant clients may need to transport it and store it carefully, in a slightly humid room which is not too dark ('diffused light'). Sungus also produces other red-skinned varieties such as Asante, Kenya Karibu and Dutch Robin (from ADC). Sungus produces small quantities of the white-skinned variety, Tigoni (from the KARI potato research centre in Tigoni). Much of Sungus's source seed is foundation seed from KARI and ADC.

Sungus perceives certification as too demanding and costly. The lack of certification from KEPHIS means that it is against the law for Sungus to sell seed from its premises. As Sungus seed enterprise is not inspected, diseases such as bacterial wilt, which may show no symptoms at high altitudes, may go undetected.

Land. Sungus has 24 hectares of farmland, and rents another 6 hectares. Seed potato is planted all year round, 2 hectares every month throughout the year. With good planning and good weather, monthly planting allows Sungus to manage with minimal seed storage.

Infrastructure and equipment. Sungus has several home stores big enough for several tonnes of seed, which are still not enough, even with staggered seed planting. It plans to expand and put up a recommended potato seed storage facility. Sungus hires machinery, especially tractors for ploughing. But Sungus is planning to buy machinery, especially for harvesting, the most expensive and labour-demanding activity.

9.4.3 Cash flow

Sungus farm is run as a professional seed business. Anne keeps the capital separate from other entities. Sungus does not attract clients through credit or offer after-sale services. However, credit of up to 20% of the price is occasionally offered to loyal clients.

Sungus prefers to 'set standard prices that encourage clients to return' and is not biased in favour of friends or kin. Sungus sells a 50 kg bag of seed potato at KSh 2000 ($27), whereas KARI sells a 50 kg bag of foundation seed for KSh 2200 ($30). Because KARI's seed is subsidized and of excellent quality, there is never enough of it to go round, which is why Sungus can charge nearly as much for their seed as foundation seed.

Sungus's most important strategy to spread cash flow is staggered planting. Other strategies include building capital reserves, learning new seed production and management skills, collaboration with research and development institutions, diversification of varieties and strict adherence to technical advice, which minimizes disease and loss. Sungus has shied away from seeking a commercial loan. Anne believes that bank loans are risky and that it would be hard to persuade a bank to give her a loan.

9.4.4 Marketing

Kenya's demand for good ware potatoes is high and rapidly increasing; Sungus's clients sell all their produce quickly and profitably. According to Sungus, high prices for ware potatoes encourage farmers to sell all their produce and not save any as seed. Seed potato storage, transportation and management are not as straightforward as with cereals or legumes. For instance, some of the popular varieties break dormancy fast and therefore require special storage conditions with diffuse light. So customers will buy seed potato even if the price increases. There is so much demand for seed that some ware potato merchants select small tubers and sell them as seed, even though they have not sprouted. Sometimes farmers can find no other seed when they need it and are forced to buy this seed or plant another crop.

Sungus Enterprise is situated in an area where many smallholders grow potatoes and is thus near many of its customers. Seed sells rapidly after harvest and having to tell customers that they have to wait until the next harvest shows Anne that they cannot satisfy all of the demand. It is increasingly serving larger-scale farmers and seed traders. Formal institutions, like secondary schools with school gardens, are 'too demanding, and often ask for post-sale services' that drain Sungus's profits or labour. For instance, secondary schools insist on physically clean seed tubers. This means Sungus has to use plenty of clean water and extra drying space.

Sungus has earned a good name through discipline and reputable management. It adapts fast and is usually the first to offer clean seed of new varieties in commercial quantities. Sungus's customers learn of new materials from the employees, from Anne herself, but mostly through her loyal clients and her network, including MoA staff, traders and researchers.

Because Sungus is unable to certify its crop, it means that it cannot brand or label its potatoes as seed tubers. It does all the packaging in plain new 50 kg bags. Marketing is through word of mouth, by customers, employees, social networks and links with researchers and institutions. Sungus is proud of the trust that it has earned by producing clean, uncertified seed.

9.5 Challenges and Strengths of the Seed Enterprises

Western Seed is challenged by its smallholder customers, who save seed rather than buying it all every year, and who cannot afford all the fertilizer that would help certified seed to reach its full yield potential. It is difficult and expensive to develop even one new variety for marginal areas, and Western Seed cannot invest in breeding varieties that customers may not buy.

Besides the usual shortcomings associated with community-based enterprises, such as capital constraints, low technical capacity, unrealistic expectations, blurred structure and dependency on aid, MUDIFESOF faces some challenges unique to soybean production, such as unfamiliarity of consumers and poor value chain development. It will have to learn to manage capital and labour on a larger scale if it is going to be a major player in the seed sector in Kenya. They will need to focus more on their brand outside their immediate locality. Their name is unfamiliar and difficult for clients to remember or understand. Clients simply referred to MUDIFESOF as 'the soybean group'. Compared with Sungus or Western Seed Company, 'MUDIFESOF' is going to be difficult to establish as a brand name. However, having a well-known brand name also makes an enterprise vulnerable to counterfeiters, such as the ones selling poor quality seed in falsified bags labelled 'Western Seed Company'. Catching counterfeiters is easier said than done.

Certification is easier for soybean than for potato. Compulsory certification for seed sales puts enterprises like Sungus at a disadvantage more than those producing seed of other crops. Community-based organizations and family seed enterprises can provide clean seed and serve smallholders, especially through partnerships with research institutes and seed companies, and should not be forced into certification.

In spite of (or perhaps because of) Kenya's advance in seed legislation, there are still inconsistencies in the legal and regulatory framework. The seed industry is governed by several acts of parliament and enforced by different institutions besides KEPHIS. This duplicates efforts and creates conflicting mandates. For example, fertilizer quality and use fall outside the KEPHIS mandate and yet they can directly or indirectly influence seed quality.

MUDIFESOF and Sungus do not have easy access to affordable credit. They see credit as risky, and they avoid debts.

A large supply of cheap, informal seed (sold as grain) drives down seed prices and discourages investment by seed enterprises, because poor customers are attracted by low prices. Electricity and construction are expensive, further curbing investment. There is also the low technical capacity of the research and government institutions, especially for potato, but also for all other crops. Available staff of government and research institutes cannot, for instance, supervise or offer useful technical follow-up for small enterprises that do not have production capacity. Much of such support is through collaborative programmes, whose efficiency is at times hampered due to ownership of material and publication issues.

Seed potato is bulky and expensive to buy and transport. Seed buyers lack seed quality information, such as the meaning of 'certified seed' and 'percentage of germination'. Limited amounts of foundation seed are available. There is also a lack of capital or affordable appropriate machinery, e.g. potato harvesting equipment.

An enterprise like Western Seed has to deal with many different agencies, e.g. for fertilizer, seed certification, export, taxes, monitoring pesticide use, and other things. They find this tedious and feel that transactions would be faster if they could deal with fewer government agencies.

References

Gildemacher, P.R., Demo, P., Barker, I., Kaguongo, W., Woldegiorgis, G., Wagoire, W.W., Wakahiu, M., Leeuwis, C. and Struik, P.C. (2009) A description of seed potato systems in Kenya, Uganda and Ethiopia. *American Journal of Potato Research* 86(5), 373–382.

Guyton, B., Sogo, F., Mogire, J. and Njuguna, R. (1994) *Kenya's Irish Potato Sub-sector Characteristics, Performance and Participants' Information Needs.* Government of Kenya Market Information System Report No. 94–01. Ministry of Agriculture, Nairobi, Kenya.

KEPHIS (2008) *Annual Report and Financial Statements July 2007–June 2008.* Kenya Plant Health Inspectorate Service (KEPHIS), Nairobi.

KEPHIS (2009) *Kenya Plant Health Inspectorate Service (KEPHIS) Newsletter.* KEPHIS, Nairobi.

Mango, N. and Hebinck, P. (2004) Cultural repertoires and socio-technical regimes: maize in Luoland. In: Wiskerke, J.S.C. and Douwe van der Ploeg, J. (eds) *Seeds of Transition. Essays on Novelty Production, Niches and Regimes in Agriculture.* Royal Van Gorcum, Assen, pp. 285–318.

Misiko, M. (2007) Fertile ground? Soil fertility management and the African smallholder. PhD thesis, Wageningen University, The Netherlands.

Organic Farmer (2009) The magazine for sustainable agriculture in Kenya. No. 51, August 2009. Biovision, Nairobi.

Republic of Kenya (2007) *Economic Survey.* Kenya National Bureau of Statistics, Nairobi.

Republic of Kenya (2009) *National Seed Industry Policy.* Ministry of Agriculture, Nairobi.

Tinsley, R.L. (2009) *Assessing the Soybean Value Chain Analysis in Kenya.* CNFA Farmer to Farmer Program, November–December. Colorado State University, Fort Collins, Colorado.

10 Uganda: Dreams of Starting a Company

Paul Van Mele, Michael A. Ugen, David Wanyama, Robert Anyang, Jean Claude Rubyogo and Louise Sperling

10.1 Introduction

10.1.1 Agriculture

Landlocked and with 16 million hectares of cultivable land (of which only a third is farmed), agriculture is the key activity for Uganda's 28 million people, contributing 43% of the country's gross domestic product and almost half of its export earnings. Coffee, cotton, tea, sugarcane and tobacco are the main cash crops. Agricultural output (cash and food crops) grew at 2.4% in 2008. But, with an annual population growth of 3.4% per year, Uganda will have twice as many mouths to feed by 2035.

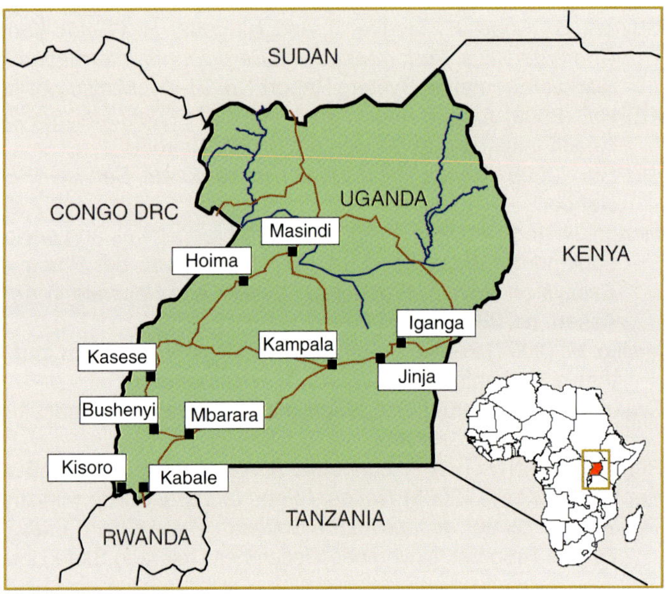

More than two-thirds of the country is a plateau, between 1000 and 2500 metres above sea level. Most of Uganda receives 1200 mm of rain spread over two seasons and produces two crops per year. These favourable conditions have turned Uganda into the food basket of the region and both crops and seeds are exported to neighbouring countries. The two rainy seasons also create challenges for the seed industry. When crops sown in March, at the beginning of the first rainy season, are harvested around June–July, farmers are already searching for seed for the second growing season, which starts in July–September. As companies are still busy harvesting and processing, this creates a general and recurring feeling of seed shortage.

When peace and stability returned to the north of Uganda in 2008, many farmers started opening up land to grow maize, beans and upland rice (only introduced in the country in early 2000). The demand for seed of these crops further increased in the south as substitution for bananas (devastated by a rapidly spreading bacterial wilt disease) and for cassava (decimated by cassava mosaic and brown streak viruses). Hospitals and boarding schools reopened, serving mainly maize and rice dishes. Other major food crops are sweet potato, cowpea, groundnut, sorghum, finger millet and potato.

10.1.2 Seed systems in Uganda

The informal sector includes NGOs, farmers' groups and family farms. There is little or no government regulation. The informal sector supplies an estimated 95% of the seed of open-pollinated food crops and practically all the planting material of vegetatively propagated crops like banana, cassava, potato and sweet potato.

The formal sector is fully regulated. The National Agricultural Research Organization (NARO) coordinates variety development of all crops, except sugarcane, tea, tobacco, flowers and most vegetables. Although responsible for producing breeder and foundation seed, it cannot do it all so it supervises companies and farmers' associations to produce foundation seed, which is then passed on to outgrowers or to other seed companies to produce certified seed, supervised by the National Seed Certification Service. Some quality seed is not grown from foundation seed, has not been officially inspected or does not fully meet the certification standards. This seed is then labelled and sold as standard seed. Some seed, such as the current sunflower hybrids, that has been tested and approved by the National Variety Release Committee is imported directly by the private sector. Commercial oil palm seedlings are also imported directly.

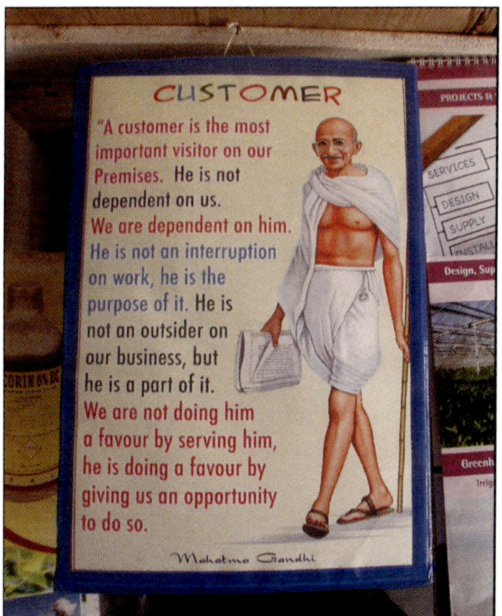

Customer service requires the right attitude, as highlighted on this poster in John Kyomya's farm supply centre in Masindi.

Seventeen national seed companies and one multinational (Monsanto) operate in the country, the latter producing about 400 tonnes of DK 8031 hybrid maize seed in Uganda in 2009. Two international companies supply seed to Uganda: Seed Co. imports hybrid maize seed from Zambia, whereas Pannar imported about 30 tonnes of hybrid maize and sunflower seed from South Africa (2009). Seed companies sell to about 50 distributors, who sell to about 2000 agro-dealers (called stockists in East Africa), who sell seed, fertilizers, agrochemicals and farm implements in rural areas.

About half of the seed companies, distributors and agro-dealers belong to the Uganda National Agro-input Dealers Association (UNADA), which provides training, market promotion and linkages. Since 2009 credit guarantee funds have allowed agro-dealers to buy enough seed. They pay 50% to the supplier and UNADA guarantees that the remaining 50% will be paid within 2 months, by which time most of the seed will have been sold to farmers.

According to the industry, farmers have become more aware of the benefits of certified seed over the past decade. When seed is accessible, Ugandan farmers are eager to try out new varieties, although until now the industry is disappointed at farmers' reluctance to buy hybrid maize (sold at about $1.60 per kg, twice the price of OPV seed). All certified seed is coated and for many farmers the coloured dressing has become a synonym for quality. This has inspired crooks to sell fake seed and has created some mistrust in commercial seed. One of the biggest challenges of the seed industry is to protect its market from fraudulent seed suppliers.

10.1.3 Evolution of the seed industry in Uganda

Before 1968 the seed sector was predominantly informal. In 1970 the government started a scheme to maintain improved varieties of all crops (except vegetatively propagated ones). It later became the Uganda Seed Project. By 1999 the responsibility for producing, processing and marketing seed was felt to be too great for a government department, so the project was transformed into a public liability company, Uganda Seeds Limited. It was privatized in 2004 and sold to FICA (Farm Inputs Care Centre) Seeds Ltd in late 2007.

The Idi Amin regime (1971–79) sparked political and civil upheavals, which damaged the seed industry's infrastructure and human capacity, and reduced the availability of improved seed to farmers. During this period, informal seed was vital. Following liberalization in 1994, private seed companies emerged and in 1999 assembled under the Uganda Seed Trade Association (USTA) to oversee the development of the industry (IFDC, 2008).

Registered companies may import and sell vegetable seed. For other crops they rely on outgrowers, often organized into associations of 100–500 members to facilitate access to credit, inputs, training, setting seed prices and marketing. Certified seed amounted to about 10,000 tonnes in 2009 (Table 10.1), accounting for about 4.5% of the seed used. Certified millet, sorghum, bean and groundnut seed dropped after 2005 when the relief market dried up.

Dealers say that there is still a large unmet demand for maize, rice and beans and seed enterprises are working round the clock to meet farmers' demand. According to Larson and Mbowa (2004), the market potential of maize alone is about 20,000 tonnes, since an estimated 8% of the area under maize is planted to hybrids and another 12% to improved OPVs. Others think hybrid maize accounts for only 1–2% of the planted area and improved OPVs for about 8%. Most farmers recycle their OPVs for too long so a lot of degeneration is evident in the field (Clive Drew, personal communication).

Many companies and seed producer groups are becoming mechanized. As Nelson Ojwiya from China HuangPai Food Machines says: 'The seed processing mentality has changed because now people start to appreciate the added value of clean seed.

Table 10.1. Seed certified (tonnes) in Uganda, 2004–2009. Source: National Seed Certification Service.

	2004	2005	2006	2007	2008	2009
OPV Maize	5,300	3,896	4,280	4,050	5,635	6,952
Hybrid maize	1,032	1,552	965	777	304	1,410
Rice	330	564	708	256	1,103	1,286
Bean	1,260	634	272	50	119	440
Sorghum	355	1,280	320	40	–	134
Millet	300	414	22	16	–	–
Soybean	30	100	158	45	40	54
Groundnut	260	135	120	15	25	8
Sesame	30	–	30	–	–	115
Sunflower	–	100	226	–	2	–
Total	8,897	8,675	7,101	5,387	7,679	10,397

Last year we installed a complete system (cleaning, treating, grading and packing) to the Bakusekamajji Women Farmers' Development Association and sold seed processing equipment to various other enterprises.' The director of the Chinese company has negotiated agreements with all major commercial banks to offer favourable loans to its clients. Some Nigerian seed companies are also using inexpensive, functional Chinese equipment (Chapter 4).

10.1.4 Seed legislation in Uganda

The Seed and Plant Act of 2006 stipulates the promotion, regulation and control of plant breeding and variety release, multiplication, conditioning, marketing, export/import and quality assurance of seeds.

The seed law encourages the private sector in research and breeding. As long as no rules exist for intellectual property rights and plant breeders' rights, breeders and companies are reluctant to invest in variety development (Danielsen, 2004). This may change in the near future since a Plant Variety Protection Bill is now before Parliament. The Seed and Plant Act also covers vegetatively propagated crops, but private seed companies have no interest in them, citing low profit margins. Apart from one private commercial tissue culture facility (AGTL), multiplication of such crops is currently being done by research institutes, NGOs, farmers' groups and individual farmers.

The Seed and Plant Act supported the creation of: (i) the National Seed Board, which advises the government on the performance of the seed industry; (ii) the

National Variety Release Committee; and (iii) the National Seed Certification Service (NSCS). With little government support and despite efforts from donors to enhance public–private collaboration in several of these bodies, many people perceive these institutions to be slow and poorly equipped to do the job. The Danish International Development Agency (DANIDA) strengthened the National Seed Certification Service, and labelling of certified seed became standard practice, which slashed the number of complaints about fake seed. In 2009, DANIDA also supported the company Chemiphar in gaining ISTA accreditation as a private seed testing laboratory.

Although field inspection of seed crops is based on OECD standards and Chemiphar is ISTA accredited, exporting seed to Kenya will still not be possible as long as KEPHIS (the Kenyan certification authority) has not carried out the field inspections.

10.1.5 The broader context shaping the national seed sector

Shortage of certified seed is compounded by the volatile demand from NGOs and relief agencies, who buy large stocks of seed from companies to distribute to conflict areas and to neighbouring countries like Rwanda, Democratic Republic of Congo and Sudan. This is unlikely to change in the near future as 'overall, the balance of (aid) assistance has shifted away from long-term development to emergency assistance' (Lele, 2009).

Soaring food prices during the 2008 food crisis stimulated farmers to buy certified seed. The increased demand resulted in more requests to register as a seed company.

Apart from higher grain prices, regional market integration has also benefited the seed sector. The East African Community (EAC), for instance, maintains an import tariff of 75% on rice. Rwanda and Burundi joined in 2006, further discouraging imports of Asian rice in the region. The tariff stimulated local rice production and had the unintended effect of boosting demand for rice seed.

In Uganda, the number of commercial bank branches drastically reduced from 1970 to 2000, with various micro-finance institutions and self-help saving and credit associations trying to fill the gap in rural areas from the early 1990s onwards (Mpuga, 2004). The Centenary Rural Development Bank is one of the few commercial banks that has been spearheading the development of farmer-friendly products and services (Seibel, 2003), although others are following suit.

SACCOs can help seed producers with their cash flow, but low-interest investment funds or bridge funds may be as important.

To improve agricultural performance, efforts to strengthen rural financial services initially focused on subsidizing credit and imposing low interest rate ceilings, but this led to dependency, poor payback, increased debt and reduced creditworthiness of the micro-finance institutions. More recently, donors have created investment funds for agrobusinesses. Uganda and Mali stand out as having successfully established effective joint donor and multi-stakeholder financial coordination systems in Africa (AMAF and WWB, 2008). Seed companies prefer such sector-wide efforts and consider donor support to individual companies as distorting. Some farmer seed-producer groups began to explore how SACCOs (saving and credit cooperatives) can support them.

The following case studies describe Uganda's first private seed company, a women's association that became a respected maize and rice seed supplier and dreams of becoming a company itself, and a farmers' group now supplying thousands of farmers in their area with quality bean seeds.

10.2 Nalweyo Seed Company (NASECO)

10.2.1 History

Apart from an abandoned cotton ginnery, Nalweyo village, 40 km south of Hoima in western Uganda, had no real infrastructure. Local people survived on tubers and bananas in cleared forest areas, grew tobacco for a company and lived in dire poverty. With support from the Belgian Survival Funds, the non-governmental organization ACT (now called TRIAS) launched a project in the mid-1990s on land owned by the Hoima Catholic Diocese. The project obtained banana suckers, cassava cuttings and seed of cereals and pulses from NARO, multiplied them and supplied farmers with free material. In 1996, 2 years after the new seed law opened the door to private investments, the project established the Nalweyo Seed Company (NASECO).

Later on, Nicolai Rodeyns, a Belgian agronomist, joined the Diocese as co-owner of NASECO, the first Ugandan private seed company.

The beginning was not easy; it was the first time in Uganda that someone was competing with the government seed company. NASECO had limited facilities and lacked market intelligence. And NARO had never worked with the private sector before. With little experience in growing seed, NASECO had to learn the tricks of the trade as it went along. Seeing more future in cereals and pulses, which were easier to handle and market, by 2004 NASECO had phased out vegetatively propagated crops.

NASECO decided to grow foundation seed to supply emerging seed companies, which kept the quantities small and manageable. But NASECO could not survive on selling just foundation seed to a few new seed companies. When in 2001 the three (at the time only) enterprises (FICA, Harvest Farm Seed and NASECO) received a grant from a USAID-funded project, NASECO decided to invest $60,000 in key farming and processing equipment to start producing certified seed.

In 2003, NASECO tested two cowpea varieties from the International Institute of Tropical Agriculture (IITA). Multiplying the seed proved such a challenge that even now the company cannot supply the requested amounts (Box 10.1).

The first experience with hybrid seed production was not easy either. When growing hybrid seed for a South African company, NASECO contracted out the processing, but, in his rush to finish the job, the contractor dried the seed at 60°

instead of 40° and killed the whole batch. NASECO decided to speed up investments in drying facilities at its own plant.

'Real disasters occur when seeds germinate in the bags. One time the pressure to deliver large amounts of seed quickly was so high that we failed to dry the seeds sufficiently. We almost had to close down our business. That was another major lesson. You should never promise what you cannot deliver without compromising quality,' recalls Nicolai.

In 2004, Monsanto contracted NASECO to grow hybrid maize seed. Through the technical advice it received, NASECO mastered professional hybrid seed production.

In 2009, it produced over 2500 tonnes of certified seed of maize (50%) and rice (20%), along with standard seed of beans (10%) and various other crops (Table 10.2). NASECO has about 30 varieties of ten crops and four hybrid maize varieties. Its flagship varieties are Longe 5 maize, Nerica 4 rice and K132 beans. Some seed, like sorghum, millet and groundnut, is sold as standard seed.

Its current strategy is to target the retail market, invest in collaborative research, establish an attractive and diverse product portfolio and raise hybrid maize production.

Box 10.1 The challenge of cowpea seed.

Farmers know what they want, but this is not always easy for seed companies.

'The old cowpea varieties Secow 1 and Secow 2 were prone to insect attack and farmers sprayed them up to seven times. As they mainly grew cowpea for the leaves, the insecticides were poisoning people's food so we persuaded the National Variety Release Committee to release the new varieties CP Kunde and CP White. They yielded 500 kg more than the old varieties. And, apart from the leaves, the beans were also better appreciated. They needed only 40 minutes to cook instead of 3 hours,' recalls Robert Anyang, at the time working with NASECO.

'But one of our outgrowers ate the leaves, liked it so much and told the entire village. Half of the leaves vanished, and seed production dropped. As if that was not enough, some also ate the fresh pods and tried out the beans. At the end we were able to collect only one bag instead of eight bags from each grower,' he says.

According to NASECO's director, 'The only option now to produce this cowpea seed seems to be to overproduce the crop in one village so that each member is entirely fed up with cowpeas and will eventually sell the seed back to the company. We know that we will receive only 50% back, but at least we will get 50% rather than nothing.'

Although initially NASECO mainly produced foundation seed and then moved into certified seed production, under an arrangement initiated by the Ugandan Seed Trade Association private companies started producing foundation seed again. In 2009 NASECO sold 10 tonnes of foundation seed of maize (Longe 5) and 5 tonnes of upland rice (Nerica 4), mostly to younger seed companies and independent growers. The company is expanding its foundation seed portfolio including beans, sorghum and soybeans.

10.2.2 Structure

Management. NASECO's vision is to work with and for farmers, building relationships and respecting deals. Profits are reinvested in fundamental needs of the company rather than in expensive cars and air-conditioned offices.

Table 10.2. Seed produced (tonnes), NASECO. Source: NASECO.

	2005	2006	2007	2008	2009
Maize OPV	864 (2)[1]	710 (3)	682 (3)	793 (2)	1,360 (2)
Maize hybrid	2 (1)	267 (2)	259 (2)	445 (3)	208 (3)
Bean	69 (2)	226 (4)	490 (5)	434 (3)	196 (3)
Rice	458 (2)	251 (4)	297 (3)	497 (3)	417 (3)
Sorghum	84 (2)	106 (2)	22 (2)	77 (2)	129 (2)
Finger millet	2 (1)	1 (1)	52 (1)	12 (3)	86 (1)
Soybean	16 (2)	17 (2)	28 (3)	99 (3)	4 (1)
Groundnut	141 (4)	55 (4)	434 (4)	244 (4)	129 (4)
Cowpea	75 (2)	72 (3)	63 (3)	81 (3)	74 (2)
Sesame	–	54 (1)	6 (1)	37 (1)	1 (1)
Sunflower	–	–	8 (1)	–	1 (1)
Green gram	–	–	66 (1)	1 (1)	23 (1)
Total	1,710 (18)	1,759 (26)	2,405 (29)	2,718 (28)	2,628 (24)

[1]The number of varieties is given in parentheses.

It believes in setting priorities and achievable targets. Choosing which varieties to produce is based on market intelligence. By producing and testing new lines and varieties on their own farm, NASECO can immediately assess how they behave, their particular needs and how farmers are likely to respond to them. Seed of open-pollinated varieties (OPVs) has been at the core of NASECO's business for the past 15 years and is likely to remain so. It considers OPV seed as a stepping stone from farm-saved seed to hybrid maize seed. The market for OPV seed is huge and NASECO believes that even the growing number of seed companies will not be able to satisfy the demand.

NASECO continues to learn, innovate and network. In 2008–2009, for instance, its director attended a seed business management course organized by CIMMYT (International Maize and Wheat Improvement Center).

Staff. Initially the company had few permanent staff and it was hard to assess the skills and incentives of job seekers. As the company built its own expertise and grew, recruiting staff with the right skills and attitudes became even more important. NASECO currently employs 30 full-time staff and over 200 day workers. Responding to advice from the National Seed Certification Service, they recruited three experienced staff to oversee field and processing activities.

Outgrowers. When the company started producing certified seed rather than just foundation seed, it developed an outgrower scheme. Starting with about 30 growers in 1999, this evolved to over 300 in 2009. All are smallholders. Foundation seed is produced on the company farm. The growers receive foundation seed, training and a guarantee that NASECO will buy back their seed at premium prices. Recently, NASECO organized the outgrowers into groups of five to 30 members with elected leaders to enforce strict seed procedures and compliance with the signed contract. Issues and targets are discussed at monthly meetings. Having outgrowers in different areas helps to spread risk. Smallholders have their own strategies to invest and cope with risks and can become large farmers if given the opportunity (Box 10.2).

Land. NASECO leases 400 hectares of land from the Hoima Catholic Diocese, its major shareholder, of which 120 hectares are under cultivation, about 7 km from the seed factory. Ever since the government improved one of the nearby roads, NASECO has maintained its surface after every rainy season. Having a good road from the farm to the processing plant is worth the investment. It saves time and limits damage to equipment.

Infrastructure. The abandoned cotton ginnery was not really set up to handle seed, but at least the old buildings gave the company a roof over its head. The first seed was manually cleaned with local labour and dried in the sun. Being in a remote location, the NASECO farm and processing plant have no access to the power grid. But the electricity supply is unreliable anyway and being independent of it makes up for the higher cost of generating its own power. About 15 years after its formation, NASECO still upgrades its equipment whenever money is available. It opened a marketing office in Kampala and another seed processing unit just outside Kampala to handle seed produced by outgrowers far from the main processing plant.

Box 10.2 The story of an outgrower.

The only thing Fred B. Magenzi brought with him was his family and his experience in growing maize. When he left his home near the Rwandan border and settled in Mumbali village in 1999, he heard that there was a company that had seed. He walked 8 km to NASECO and proposed becoming an outgrower.

He first grew maize seed on rented land and gradually bought his own. In 2002, he planted 0.4 hectare of the recently introduced new rice variety, Nerica 4. In 2005, he planted 2.4 hectares, but during the harvest a hail storm destroyed everything. The 1.2 hectare he planted the next year was destroyed by drought and he and his neighbours lost confidence in growing rice seed.

All along he has continued to grow maize seed. He now owns 48 hectares of land, 8 of which he has planted with eucalyptus trees. All his children go to school; he has a house with an iron roof, 19 cows and 24 sheep. Having learned from his skills, some of his neighbours have also become outgrowers for NASECO.

Equipment. Although it started small, using animal traction, NASECO is now better equipped and has increased the area planted, especially for cereal seed. Legume crops are shorter and have to be cut about 10 cm above the ground level, making machine harvesting difficult on uneven fields, so less land is devoted to these crops. In the near future, the company intends to invest in land preparation to expand the area and mechanize the seed production of legume crops.

All processing takes place at the plant: preparing the seed, cleaning, drying, storing, seed conditioning and packaging. NASECO continuously assesses the performance of its equipment and innovates whenever needed. Drying and storing are crucial in a country with two rainy seasons. NASECO soon replaced solar drying with a flatbed dryer, but it dried unevenly so NASECO changed to a batch dryer. Later on, two recirculating dryers were added, each with a capacity to process 15 tonnes of seed within 8 to 10 hours.

NASECO built its dryers to be fuelled by maize cobs. Nicolai is a hands-on man, not the type who manages his company from behind a desk. He is convinced that part of the reason for the success of his company is its paced growth and that it has resisted the temptation to buy expensive equipment with risky bank loans.

Now that it has solved the problems of drying and storing, NASECO intends to invest part of its profits in a fully automated packing line before the end of 2010. This is crucial to guarantee quality and avoid fraud, both key to maintaining customer trust.

Batch dryers in NASECO's seed processing unit.

Recirculating dryer fuelled by shelled maize cobs.

Quality control. NASECO and the seed certifying agents train outgrowers to ensure quality production. The agents also inspect the fields at least twice for OPVs. NASECO conducts basic seed tests (germination and moisture content), but can also submit samples to the National Seed Certification Service before buying seed from outgrowers in case of doubt. The certification agents take samples before issuing official labels.

10.2.3 Cash flow

NASECO learned that growth has to be paced: walk before you run. The company relied as little as possible on bank loans. Many new seed companies want to do too many things at once and often can't manage it, resort to bank loans, make serious mistakes and go broke.

NASECO does not fear competition, even though it competes on an uneven playing field (e.g. some companies receive large donor grants). Its confidence is based on sound business management, experience, unmet national and regional demands and its continued innovation with new seed technologies. A key challenge is managing its cash flow. Fortunately, Uganda's double cropping seasons reduce cash flow needs. While NASECO has built up relations with many distributors and agro-dealers, it is important to know who is creditworthy and who is not. One has to constantly monitor cash flow by balancing suppliers' needs and customers' credits.

Apart from seed, NASECO does not provide any other inputs on credit to its outgrowers. It pays them a premium price of 50% above the grain market price at harvest time, rather than holding back payment until the seed is sold. This farmer-friendly approach has created a lot of goodwill and neighbours of outgrowers have also become interested.

Many seed enterprises, NGOs and development agencies believe that seed has to be cheap to be affordable to farmers. NASECO believes that affordability is not just a matter of price, but also includes proximity, quality and quantity. Offering quality seed in smaller quantities may indeed make it more affordable and has become a key investment priority for the company since 2010.

10.2.4 Marketing

While most seed companies in Uganda target NGOs, right from the start NASECO has worked with and for farmers. In 2009, about 10% of NASECO's production was sold to the government, 30% to NGOs and 60% to the retail market (most of which is then sold to farmers) (Table 10.3). 'The easy thing about selling to NGOs is that they often place larger orders at a time and that they pay within a month or so. But they also frequently change interventions, and chances are high that they will renounce contracts, making planning difficult. Most of them also look more at the price rather than at quality. When there is a tender, some companies buy up grain, clean it and offer it cheaply, so there is no way you can compete with them if you take quality seriously. Therefore we want to focus more on the retail market,' says Nicolai.

Table 10.3. Clients of NASECO.

	>1996	2000	2009	2015 (predicted)
Agro-dealers	3	1	1	3
Distributors	1	3	2	1
NAADS/Local government	–	–	3	4
Individual farmers	–	5	4	7
International relief agencies	–	2	5	5
Local relief agencies	–	4	6	6
Commercial farmers > 100 acres	–	–	7	2
Companies	2	6	8	8

Ranking assessment by senior management of seed enterprise, 1 being the most important.

The retail market has its own complexities. The peak sales period is at planting time, in March and July. To ensure farmers' access, NASECO tries to have its seed in the shops year-round. At first, NASECO built on the network of agents established by the Uganda Seed Project. For instance, Uganda's largest seed dealer, El-Shaddai, buys seed from all major seed companies and has a network of over 60 agro-dealers across the country (Box 10.3). In 2009 El-Shaddai sold about 500 tonnes of certified seed, of which more than 70% was non-hybrid.

Box 10.3 El-Shaddai International, Uganda's largest seed distributor.

Having been in the business for about 15 years, Patrick Makwetta, owner of El-Shaddai, has seen many changes in the seed sector. 'Demand is increasing every year. Many new seed companies are too ambitious and try to look for funds. As many of them may deliver substandard products this may lead to total confusion,' says Patrick. 'It would be better if they would focus on a few varieties only, at least in the beginning.'

Asked what donors could do to support the sector, he says: 'Seed dealers are hostages of what companies offer, so companies should be supported in doing R&D on new varieties and in producing breeder seed. Young enterprises that already have access to land could be supported by investments in infrastructure. Donors could help to provide cheap finance by supporting banks to develop favourable products, but they should never give away free money as this distorts the market.'

John Kyomya has a farm supply centre in Masindi and is a key distributor in western Uganda, supplying local agro-dealers and serving about 3000 farmers, of which at least 600 buy seed every season. He plants all the varieties he sells, including the ones he buys from FICA and NASECO. 'Every new variety we grow for consumption first: test how it behaves, organize demos and then cook the rice or roast the maize and eat it with other farmers. As we need to give advice to our clients, we need to know what we sell and there is no better way than growing it yourself,' John says. In 2009 he sold 11 tonnes of Longe 4, 9 tonnes of Longe 5 (a maize with more proteins) and 9 tonnes of Nerica 4 (an upland rice variety developed by AfricaRice). Half of John's stock went to the National Agricultural Advisory Services (NAADS) and the other half to farmers via agro-dealers. In the past few years many new agro-dealers have joined the fray, or switched tactics. For instance, Nyati Mills in Hoima came up with the idea of establishing an agro-input shop next to their rice mill. They bought certified rice seed from NASECO via John Kyomya. Sales increased from 1.5 tonnes in 2007 to 3.8 tonnes in 2009.

NASECO buys printed bags that contain company and variety information and have a line with rainbow colours, which makes them hard to copy. The bags are transparent so that the clients can see the seed. While hybrids are packed in 2 kg sacks, seed of non-hybrids is packed in 5, 10 or 15 kg bags depending on the request of the agro-dealers. When farmers ask for smaller quantities, seed bags are opened, placing farmers at the mercy of the agro-dealer's honesty. NASECO wants to protect its good reputation by producing smaller packs of 1 kg and even 0.5 kg.

NASECO invests in transparent bags so that the clients can see the seed. The bags have a line with rainbow colours, which makes them hard to copy.

When official quality control is insufficient, innovating with packaging and product development helps protect the company's name and farmers' confidence in certified seed.

Testing new varieties also helps to keep the lead. NASECO does adaptive research in maize, rice, sorghum and sunflower. For instance, the red and white sorghum varieties now grown by farmers were released 15 years ago and Nicolai feels that the time is right for some new varieties.

NASECO offers farmers several good varieties every season. At first, clients came by themselves to look for seed, but, as the business grew, NASECO invested more in marketing, which itself was seen as an ongoing experiment by its director. The focus is not to grab clients of other companies but to create new customers (e.g. by switching from home-saved seed to improved seed) through demonstrations, broader networks, improved availability and affordability.

10.3 Bakusekamajja Women Farmers' Development Association

10.3.1 History

The rolling hills at the source of the Nile in eastern Uganda have attracted foreign investors for a long time. Tea and sugarcane plantations are etched into the landscape. Apart from rice in the lower, often swampy areas, farmers in and around Iganga district grow maize and rice as cash and food crops, along with several other subsistence crops, such as sweet potato and beans.

Women's associations often develop multiple enterprises, of which seed may be one.

In 1986, 16 women decided to establish the Bakusekamajja Women Farmers' Development Association to ban hunger from their village near Jinja. With music, dance and drama they sensitized women to join hands, be more active in agriculture and start growing fruit like papaya and passion fruit. The local church frequently hired them to perform, and over the years they became more widely known for their commitment to improving food security.

Eastern Uganda is a region that has been swamped with projects for the past 15 years (30 seasons) demonstrating the benefits of improved seed, and other efforts to strengthen commodity value chains (linkage to output marketing, rural agro-dealers, etc.). Examples of initiatives include the USAID-funded Investment in Developing Export Agriculture (IDEA) and Agricultural Productivity Enhancement Programme (APEP) projects, Title II, SG2000 and NAADS. Most farmers have been repeatedly exposed to certified seed through farmer field demonstrations and learnt about the benefits.

When in 2000 NARO looked for organized groups in Iganga to test new varieties, the county suggested they contact their women's association, which had grown to several hundred members and was eager to get seed of new varieties. The old landraces were yielding poorly and improved seed was hard to get. NARO identified six women and taught them how to produce maize seed at their station in Namulonge. Soon 14 trained women, two from each zone, acted as farmer trainers. The women's association became NARO's principal group to multiply foundation seed.

In 2004 they grew 146 tonnes of maize seed (Longe 1 and 4), which largely surpassed the needs of its members, and became outgrowers selling to Victoria Seeds, among other companies.

In 2005, the national excitement surrounding upland rice (Nerica) had aroused the women's interest and they approached NARO. As a favourite of NARO, they quickly received free training and pre-basic seed. For nearly a decade now, the association has built up experience growing and marketing seed of maize and more recently also rice, perhaps initially at the expense of food security (Box 10.4).

Box 10.4 Growing Nerica seed at the expense of food security?

According to a report by CABI (who supported the Bakusekamajja Women Farmers' Development Association) rice seed production was only profitable for ten of the 21 farmers trained (Page, 2008). Growing rice and rice seed demands labour, especially during land preparation, weeding and harvesting, so farmers no longer had time to grow their maize crop.

However, by 2009, 17 out of the 21 farmers trained were making a profit and all were able to undertake both rice and maize cultivation. So what had happened? During the first year land had to be cleared; farmers were unfamiliar with best practices and had to weed four times. Soon they learned that it was better to prepare the land in December and plough once more before sowing. This killed the tough weeds and reduced weeding to two times only.

Assessing impacts before projects have advanced may give misleading results. Farmer groups need time to mature and time to experiment with new technology.

The seed growers of the Association stuck to maize and rice (Table 10.4). In 2007 they produced mainly the new maize variety Longe 5. By then all farmers had heard about it, as well as Nerica 4 seed. Charles Operemo, production officer of Victoria Seeds, said that 'there has been a strong promotional campaign for upland rice by our

Table 10.4. Seed produced (tonnes), Bakusekamajja Women Farmers' Development Association.

	2004	2005	2006	2007	2008	2009
Maize	146 (2)[1]	428 (1)	287 (1)	121 (1)	131 (1)	133 (2)
Rice	–	69 (1)	66 (1)	81 (1)	71 (1)	140 (1)
Total	146 (2)	497 (2)	353 (2)	202 (2)	201 (2)	273 (3)

[1]The number of varieties is given in parentheses.

government and the FAO to eradicate poverty, so market saturation hasn't been a big problem. Despite Nerica 4 being around for a decade, people appreciate its attributes, like good aroma and non-shattering, and that is why the sales of Nerica 4 seed never came down.' In the near future the women's association intends to embrace bean seed production.

Although the seed growers improved their living conditions, they became increasingly unhappy. Disturbed by the fact that retailers sell their seed at three or four times the amount the farmers get (and unaware of the actual costs of processing and marketing), they felt unfairly treated and in 2009 decided to try registering themselves as a company.

10.3.2 Structure

Management. The Bakusekamajja Women Farmers' Development Association has two key activities: seed growing and rearing animals through zero grazing. Only women are allowed to rear animals but men can sign up to grow seed. But men never have voting rights. Grace Bakaira is chairperson of the association and there is one cashier. Every year the female members elect a zonal leader, who monitors the activities of about 70 members. The association has 516 members in six communities. About 350 of them grow maize seed (85% women) and 407 produce rice seed (32% women). As women were already producing maize seed, working many hours a day, when rice arrived men came forward to engage in the rice seed business, attracted by the profitability.

'Although we started off with women only, men cried to join as they felt left behind. As fields could be up to 10km away from their home and men could more easily ride their bicycle there, we decided to include men for seed production,' explains Grace. The association has a family-friendly approach. During training sessions held on the weekend women are invited to bring their children along. If the husband is interested in the subject he can join as well. One of the association's by-laws for seed producers stipulates that children should never be mobilized to scare away birds from the crop, as this would keep them from attending school (Box 10.5).

Box 10.5 Scaring birds in a silent way.

Semu Kasagoya and his wife joined the Bakusekamajja Women Farmers' Development Association in 2005 and started growing maize seed (Longe 5) and later rice (Nerica 4). Seed growing has become their major enterprise.

When asked whether he faces problems with birds, like rice growers in many countries do, Semu Kasagoya sits back and starts explaining: 'We plant seed of both these new varieties when the first rains arrive and they will ripen at the same time. Birds like maize more than rice, but the damage they do to maize is limited because there is so much, so I don't really mind. But, if they have only rice to feed on, they can finish the entire crop. By the time the birds realize that there is also a rice field nearby, the rice is nearly ready to harvest. I have to spend only about 2 weeks in the field guarding my crop.'

When asked what he does during those 2 weeks, he smiled and continued: 'You must definitely not make any noise, because very soon the birds will get used to it and when they do not hear any noise they know the boss is away. It is better to stay quiet and throw a stone at them every now and then. It is the same if you want to keep thieves out of your field: you do not go shouting all day, because the day you don't shout the thief will know that you are not there.'

Land. Fields of new members are first inspected for suitability, to see if there are other varieties grown nearby. If neighbouring farmers grow the same crop, their relationship with the potential seed grower is assessed as they will need to agree on planting time or planting the same variety as the seed grown. Growers should have a plot of at least 1.2 hectare for seed production.

Equipment. The association has received basic equipment from various donors. In 2008 they bought 1.2 hectare of land, on which a year later an impressive seed processing plant was built by a Chinese company with funds from Lutheran World Relief. From an earlier project they had received a manual seed dresser.

Contracts. Each season growers decide on the variety, in consultation with the committee, after which they sign a contract with the association and receive foundation seed. Failure to sell seed back to the association can result in legal prosecution. Although several neighbours of seed growers have joined the association, many are afraid to become seed growers because of the stringent conditions set.

The committee of the association initially approached the companies to sell their seed, but since they became better known it is the other way round. 'When the companies visit us, we look at what they want and what we have on offer. We sign a contract stipulating amounts and varieties to be delivered, but we do not decide on the price. As this fluctuates too much, both parties believe it is better to negotiate that at the time of harvest,' says Grace.

Links. Apart from providing breeder seed seasonally, NARO has also played a crucial role in linking the association to companies and the seed regulatory authority. Having outgrowers well organized in an association is convenient for the seed companies, who cover the costs of the official field inspections. When the Bakusekamajja Women Farmers' Development Association becomes a registered company they will have to pay for the certification themselves.

Quality control. Despite the cost, which is likely to increase significantly in the near future, chairperson Grace is convinced that official inspection is needed as long as most Ugandan farmers are illiterate. 'Without control, selling fake seed would be open: at least now they fear,' she says.

10.3.3 Cash flow

In the past, the women's association benefited from support from the Rockefeller Foundation, CIMMYT and CABI, who trained its members on maize and rice seed production and handling and provided basic equipment. CABI developed a training manual and radio programmes.

'When farmers told us their sad stories of how they had been cheated in the past with false promises, we decided to do everything possible to pay our members at the time that we buy their seed,' explains Grace; 'it is one of the reasons that our association still exists.' Its seven-woman executive committee ensures every season that money is paid to the seed growers at harvest. The committee visits the different seed companies, negotiates seed delivery contracts and requests upfront payments of 70% at harvest time, once the price has been agreed upon. The association refrains from giving loans to its members, apart from seed loans, which are deducted at harvest time. Outgrowers receive a premium of 40–50% above the grain market price.

With no interest to keep on supplying other seed companies once they are fully registered themselves and realizing that a strong capital base is crucial to operate a seed company, they established a saving and credit cooperative (SACCO) in mid-2009. Contrary to most SACCOs in Uganda, this was a community rather than a government initiative. Six months later, members had already deposited 7.3 million Ugandan shillings ($3750). Knowing that transparency is key to survival, the SACCO set up a separate account with different signatories from those of the association. The SACCO gives loans to farmers, but only for agriculture and school fees.

With the large capital investment from Lutheran World Relief for the seed processing plant, the association signed a contract with them that it will recruit a management team and be fully operational by 2012.

The association does not fear competition, since, once it has been registered, it will be the only company operating entirely in eastern Uganda near its clients.

10.3.4 Marketing

'Our clients return to us because we deliver good quality seed that has a high germination percentage. We try to limit sales to individuals. If a farmer comes in the name of a group and buys a large amount, we charge 1500 shillings ($0.75) per kg of maize seed; individual members of our association pay 1700 shillings ($0.85), whereas non-members pay 2000 shillings ($1.00),' says Grace Bakaira. Members of the association are key clients, although, once registered as a company, the association intends to supply most of its seed to agro-dealers (Table 10.5).

Seed enterprises gain a lot by knowing the grain market. As Grace explains: 'Our main seeds grown are likely to remain the same in the near future, although demand may fluctuate. The quality protein maize, Longe 5, is really popular among schools, for instance. And, although Nerica 4 is the most popular upland variety, Nerica 10 with its bold grains mixes well with our Super landrace, which has aroma and similar cooking qualities. Traders like this as they can mix them and therefore they are on the lookout to buy grain of Nerica 10 from farmers. When there is a good market for the grain, there is a good market for the seed.'

Table 10.5. Clients of the Bakusekamajja Women Farmers' Development Association.

	2003	2009	2015 (predicted)
Companies	–	1	–
Members	1	2	3
Schools	–	3	5
Women and youth groups	–	4	4
Individual farmers	–	5	7
Prison	–	6	6
NAADS groups	–	7	8
Agro-dealers	–	–	1
Relief agencies	–	–	2

Ranking assessment by senior management of seed enterprise, 1 being the most important.

The women's association organized a series of radio programmes on rice and seed cultivation. Soon after, they received many new orders for seed and 63 new members signed up. With growing demand for support, they also established another branch, called Bakusemajja 2, with 79 members. A similar experience took place in Mali, where farmers announcing local seed fairs on a rural radio station created such a demand for quality seed that it actually led to the establishment of a seed producers' cooperative (Section 5.5.1).

Since 2002 the association has had continued access to breeder seed and developed expertise in growing and selling their seed to members and other clients. Being closely linked to NARO and conducting varietal performance trials with them, they may access upcoming varieties, unless NARO signs exclusivity contracts with other companies.

However, managing a registered seed company and a sophisticated processing plant will be a completely different ball game. Monitoring, inspection, processing and marketing will come at a cost and price of seed will inevitably go up. What the association has always resented about seed companies, namely their high seed price, may be coming back to them. Whether the investment provided by the donor will bring benefits to their association and other farmers in the region remains to be seen.

10.4 Nyamabale Bean Seed Producers' Association

10.4.1 History

The Nyamabale Bean Seed Producers' Association started as a farmer field school group in 1997. Located 30 km from Kabale town in south-western Uganda, the group collaborated with NARO and the International Centre for Tropical Agriculture (CIAT) to test integrated bean root rot management and resistant bean genotypes.

By 1995, bean production in the highlands of south-western Uganda had plummeted by 80% due to a complex of root rot diseases (caused by soil-borne pathogens *Pythium* spp., *Fusarium* spp. and *Rhizoctonia solani*) (Opio, 1999). Families were going hungry, and children and pregnant mothers were especially at risk. Resistant bean varieties were urgently needed.

Drawing on beans bred in Uganda and at CIAT headquarters in Colombia, by 1999 the Nyamabale farmer group started evaluating over 100 lines with NARO, CIAT, local traders and local extension services. By the third year they had selected the three most resistant and appreciated lines: SUG 31 (NABE 12C), which is a climber bean, and two bush beans, RWR 1946 (NABE 13) and RWR 2075 (NABE 14) (Table 10.6).

Although farmers had selected the three most promising varieties, seeds were not available. NABE 12C was officially released in 2003 but no seed company was interested due to the high labour demand for staking climbing beans. The varieties NABE 13 and NABE 14 were released in 2006, but again, despite the high demand, no seed company was interested because they were mainly suitable for the highlands, filling a narrow ecological niche.

Apart from evaluating bean varieties, the Nyamabale farmer field school also taught their members about integrated pest and disease management and integrated soil fertility management to tackle bean root rot and bean stem maggot (which usually appear together and can cause 100% yield loss). Its field school became a model and soon it started training other groups in bean production for payment.

If the highland farming communities were to benefit from these new varieties, then the group would have to produce its own seed, which it did with technical advice and support from NARO and CIAT. In 2005, the Nyamabale Bean Seed Producers registered themselves with the sub-county and district governments as producers of quality standard bean seed. The group also registered with NAADS as a community service provider to train other farmer groups.

It started small, producing 5.3 tonnes in 2005, steadily increasing its capacities by contracting other farmers to produce seeds under its direct supervision. By 2009 it was producing over 15 tonnes of standard seed (Table 10.7).

Table 10.6. Bean varieties selected by Nyamabale farmers.

Variety			Other attributes	
Official name	Farmers' name and its meaning	Yield (tonne/ha)	Biological	Consumer qualities
NABE 12C	*Musingiriro* (means a climber in Mufumbira language) and others call it sugar since they say it is sweet.	2.5–3.5	Tolerance to root rot, large attractive pods and seeds, high yield, more heat tolerant, suitable for areas with limited land	Highly marketable seeds, cooks fast, tasty and swells on cooking, tender leaves as vegetable
NABE 13	*Mulwanisa* (means it fights root rot resulting in greater availability of beans)	1.5–2.5	Resistance to root rot, large red seeds, high yield	Large red seed suitable to sell and to eat at home
NABE 14	*Muzawura* (means 'saviour' as it fought root rot disease and had more selling/ high market demand and brought people out of poverty)	1.5–2.5	Resistance to root rot, large red seeds, high yield	Large red seed suitable to sell and to eat at home

Table 10.7. Seed supplied (tonnes) by the Nyamabale Association.

	2004	2005	2006	2007	2008	2009
Bush beans	0.2 (1)[1]	1.1 (2)	1.6 (2)	2.2 (2)	2.3 (2)	1.8 (1)
Climber beans	0.3 (1)	4.2 (1)	4.9 (1)	6.6 (1)	7.9 (1)	13.8 (1)

[1]The number of varieties is given in parentheses.

Lessons learned from the community-based seed production in Kabale helped NARO to successfully scale up the system to other districts of Bushenyi, Kisoro, Kabarole, Mbarara and Kamwenge. Bean traders also started selling bean seed (Box 10.6). Formal and community-based seed systems target different markets and supply different bean varieties, resulting in complementary effects and outreach to especially hard-to-reach communities in remote areas.

Box 10.6 Making money with bean seeds.

Patrick Monday, a bean trader from Kabale district, was surprised to realize that he had never thought of earning much money from selling seeds, but, with access to the root rot-resistant, high yielding bean variety NABE 14, he is generating a good income for his family.

The seed is in high demand in communities with more land which can afford to grow bush beans. The new bean varieties fetch higher prices and more people are now growing improved seeds in Kabale.

10.4.2 Structure

Management. The Nyamabale Bean Seed Producers' Association has the vision of producing high quality seed of both climbing beans and bush beans and of working closely with farmers, NGOs, researchers and others to supply improved bean varieties to farmers at a relatively low cost, without compromising quality. It hopes to continue growing and become a seed company.

It is governed by a constitution made by the group with help from NARO and community leaders. Every 2 years it selects a chairperson, vice chairperson, treasurer, vice treasurer and secretary. Although Nyamabale group has 18 registered members, of which six are women, they work with 27 other groups totalling 470 farmers, who produce bean seeds under their supervision.

Some also grow seed potato, mostly on a small scale, for selling to communities. They get improved seeds from the Uganda National Potato Producers' Association based in Kabale and to a lesser extent from the Kachwekano Zonal Agricutural Research and Development Institute (ZARDI).

Over the years, NAADS has often contracted them to support other groups on basic crop production and pest, disease and soil fertility management. The cash earned from such services is invested in the seed business or other agro-enterprises.

Land. Land tenure is a challenge for the association. For a long time, the group depended on land donated by the other groups. After raising some funds, they bought 7 hectares of fields. They also rent land each season from other farmers. This is the most densely populated part of Uganda (327 inhabitants per km^2) with average farm size between 0.2 and 0.8 ha; climbing beans (NABE 12C) are particularly important in this region because they yield 2.0–2.5 tonnes of seeds per ha, which is three to four times more than bush beans.

Infrastructure. The association has an office, an information centre (that has reading materials and photographs on bean production, protection and seed processing) and a big, new, modern store which can hold 40 tonnes of seed.

Equipment. The association does not have large machines but performs most work by hand. It uses a drum to mix chemicals (Actellic Super dust, used to control storage pests in grain, not harmful to consumers) for batches of 100 kg of seeds. No dyes are used in seed dressing. It also has a scale that weighs up to 500 kg. Another machine verifies that moisture level before storage is at 12–15%. Seeds can stay in storage for up to 2 months before being sold.

Links. The group receives foundation seed from NARO and maintains good relations with local government, researchers, extension, NGOs and community-based organizations. Participatory varietal selection and on-farm demonstrations have enhanced farmers' and the community's knowledge about new varieties, allowing them to select varieties they prefer, and also helped growers to produce what farmers demand. NGOs, community-based organizations, agro-dealers and farmers are the key clients. The group can access government help by working closely with NAADS and other local government projects that operate in the area. Developing and managing effective partnership have been essential for system development and sustainability (Rubyogo *et al.*, 2010).

Training. NARO, CIAT and other development agencies have trained the association in seed production and postharvest management, group visioning, leadership, negotiation and communication, conflict management, business planning, farming as a business, marketing, market chain analysis, and cost and benefit analysis. Farmer groups have learned to manage their resources better and participation of both men and women has created awareness of gender equity.

Quality control. At times, seed production fields are inspected by the National Seed Certification Service, but certification has been a problem for most farmer groups. The few certifying agents cannot visit all seed producers across the country; hence seed is sold as standard seed.

Farmer groups have received training in internal quality management. A handbook developed by CIAT shows simple methods farmers can use to test seed to improve quality (David, 1998). Farmers test germination, moisture and other quality aspects. Germination tests before storage should give at least 85% germination. Initially farmers were reluctant to rogue off-types, which they thought was destroying their beans, and preferred allowing these to grow and harvesting the seeds for food. This challenge was gradually overcome.

10.4.3 Cash flow

The group agrees on a price with buyers before harvest, or may even sell seed under contract with the prices set beforehand, especially to NGOs, the Kabale District Farmers' Association and community-based organizations. Usually seeds are not stored for long as demand is high. This reduces storage costs and losses.

It has not been very easy for the group to obtain loans due to the high interest rates charged by micro-finance institutions. Other challenges include: drought and erratic weather; sensitizing and convincing communities about the advantages of quality seed (inspected rather than just grains); inadequate business planning and record keeping; seed producers' limited capital; and lack of access to formal credit.

10.4.4 Marketing

The association sells most of its seed to traders (the seed/grain traders who work the normal markets), farmers and even agro-dealers as far away as Kabale town (30 km), who normally only sell certified seed (Table 10.8). If they get there first, NGOs may take most of the seeds while in other seasons NAADS takes the most. NGOs, NAADS, traders and agrodealers take up to 80% of the seeds that are sold locally.

The agro-dealers selling bean seed are mostly members of UNADA. NARO gave them posters on bean varieties and trained them on seed handling, business management and communication to clients. Both bean seed producers and agro-dealers make announcements on the local radio in Kabale to inform potential clients about seed availability and prices.

Table 10.8. Clients of the Nyamabale Association.

	2004	2009	2015 (predicted)
Traders	5	1	2
Agro-dealers	–	2	1
Members	1	3	3
NAADS groups	–	4	7
NGOs and CBOs	2	5	6
Individual farmers	3	6	5
Farmer groups	4	7	4
Seed companies	–	–	8
Relief agencies	–	–	9

Ranking assessment by senior management of seed enterprise, 1 being the most important.

High quality keeps the customers satisfied, even at a price about double the grain price. The group sells its inspected seed at 1800–2000 shillings per kg ($0.9–1.0) while the price per kg of grain is 1000–1200 shillings ($0.5–0.6) and of certified seed 2500–3000 shillings ($1.2–1.5). While NGOs, NAADS and agro-dealers buy in bulk, the group uses small packs (0.5 kg, 1 kg and 2 kg) for farmers. The new varieties are now grown in over ten districts and have spread to Kenya and Burundi.

Bean seed produced by the association for sale in small packs at the Royal Agro-input Suppliers shop in Kabale, one of many rural shops that are members of the Uganda National Agro-input Dealers' Association.

Farmers appreciate that seed is available locally and from a trusted source that can be held accountable. The producer group has built a good reputation with NGOs and other institutions within the district and beyond for its supply of good quality seeds. Traders around Nyamabale village have been buying seed from the group and selling them around Kasese (400 km away) and as far as Rwanda and Congo (DRC). Traders say they actually have clients waiting for deliveries.

10.5 Challenges and Strengths of the Seed Enterprises

Following the 2008 food crisis, demand for seed increased and many farmers' associations, agro-dealers and business people believed that they could run a seed company. Both of the farmers' associations in this chapter were struggling to become registered, as it proved hard to fulfil all the stipulations, such as having the technical expertise to monitor the members' production. However, despite the growing number of companies, the government has not invested in quality control. Rather, it cut the staff at the National Seed Certification Service from eight to six in 2010. Being unable to cope with the already heavy task, the credibility of formal seed is at stake. Only companies that are willing to invest in quality, innovate with packaging to avoid fraud and actively market their products are likely to survive.

Also, there is little or no foundation seed available for most food security crops such as sorghum, millet, beans, cowpea and groundnut, so seed enterprises often sell standard seed rather than certified seed of these crops (Section 10.4).

National companies do not fear international or multinational seed companies taking over the Ugandan seed market, because quantities of hybrid seed are minimal. Also, Uganda's maize consumers prefer local white maize varieties for their unique flour qualities for baking bread and for brewing.

Uganda has no system of plant breeder rights, which some argue discourages investment in research (Chapter 2). However, according to NASECO, this is not a major problem: while it invests in research in new varieties it starts bulking seed and, by the time the variety is released, enters the market at least 2 years ahead of its competitors.

NASECO's main reasons for success are its visionary, pragmatic management, its rationally paced growth with a clear focus and avoiding loans, as well as its perseverance and learning from mistakes (Section 10.2). It has invested in its staff and experimented with equipment, self-pollinated varieties, seed production, processing, packaging and marketing. Most of all, NASECO developed an abiding respect for farmers, both as outgrowers and as clients. It is a company with ethical principles. With risk increasing as the company expands, managing its growth will be key to survival.

As the Uganda seed sector depends on its outgrowers, the viability of seed enterprises is linked to fluctuations in the labour market, including the opportunity costs to outgrowers of producing seed. Seed production is labour-intensive, especially for land preparation, weeding, roguing off-types and harvesting. During the first growing season, after the New Year, families have often spent their money and have school fees to pay, so farmers are on the lookout to sell their labour. During the second season, farmers spend more time on their own farm and labour becomes more expensive. There are also regional differences. In Iganga, eastern Uganda, for instance, labour is more expensive than elsewhere, because of the high migration rate to urban areas. Farmers' associations and companies operating in eastern Uganda will need to pay more or help outgrowers save labour.

The Bakusekamajja Women Farmers' Development Association has excellent links with national researchers and was able to cleverly expand its farmer customer base through educational radio programmes dealing with the crops whose seed they produce and sell (Section 10.3). Although they have been testing varieties and producing seed for a decade, they became increasingly frustrated selling to seed companies, believing that seed has to be cheap for farmers (an opinion often shared by donors and development agencies). Having received much capital investment from donors and trying to become a seed company, they will soon face the realities of running a business themselves. If they are not happy running their own company or if it proves too much for them, they can always go back to being outgrowers.

Beans are an important national crop, but when varieties have a narrow ecological niche or are difficult to mechanize, such as the climber beans in south-western Uganda, community-based seed production may be the only way, as shown by the Nyamabale Bean Seed Producers' Association. Community groups are succeeding in meeting a need (and making money), even though demand is too limited to attract the interest of any companies.

Women of both seed producing associations have actively engaged with the media, which has helped to establish their reputation and expand their membership. Based on a study in Mali, Uganda and Zambia, the International Women's Media Foundation (IWMF) found that just 4% of all media coverage was devoted to agriculture, of which rural women featured in just 7% of the stories (despite them producing 70% of sub-Saharan Africa's food) (IWMF, 2009). The examples in our book show the importance of media in establishing rural enterprises.

Considering that the private seed sector is still relatively small and young, Kiwanuka and Kintu (2004) called for policies to be inclusive of farmers' rights to keep on producing, saving and selling their own seed. They also proposed that communities should be supported by providing technical knowledge on seed saving and storage, which is one of the areas where media can play a crucial role (Van Mele *et al.*, 2007).

Farmer seed producer groups face many challenges, including labour-intensive manual harvesting and threshing, transport from the field, limited drying, storage and packaging facilities, and financial management. SACCOs may bring some credit relief, but good governance, capitalization and attractive financial services are crucial to surviving in a competitive financial market. This is exactly the support that the Uganda Cooperative Savings and Credit Union (UCSCU) gives to its members. Bridge funds and loan guarantees to commercial banks backed by the government and donors may be just as important, as also shown in Mali (Chapter 5).

Uganda is said to be one of the most entrepreneurial countries in the world. However, about 30% of the entrepreneurs shut down their businesses within the first 12 months of operation (Rooks *et al.*, 2009). Although seed enterprises are particularly tricky to establish (Chapter 2), in Uganda none of the registered ones have gone bankrupt so far. As Uganda's seed industry is still heavily dependent on non-commercial sales to government and donor-funded projects, this is the basis on which over half of the current seed companies are able to survive (Clive Drew, personal communication). In a country where many dream of setting up their own business, at least some dreams do come true.

References

AMAF and WWB (2008) *Diagnostic to Action: Microfinance in Africa.* Africa Microfinance Action Forum (AMAF) and Women's World Banking (WWB), New York.

Danielsen, S. (2004) *Scoping Mission to Explore the Opportunities for Support to the Ugandan Seed Sector under ASPSII.* Danida, Kampala.

David, S. (1998) *Producing Bean Seed: Handbooks for Small-scale Bean Seed Producers.* Network on Bean Research in Africa, Occasional Publication Series No. 29. CIAT, Kampala.

IFDC (2008) *An Action Plan for Developing Agricultural Input Markets in Uganda.* International Center for Soil Fertility and Agricultural Development, Muscle Shoals, Alabama.

IWMF (2009) *Sowing the Seeds. A Study of Media Coverage of Agriculture and Women in the Agricultural Sector in Three African Countries: Mali, Uganda and Zambia.* International Women's Media Foundation (IWMF), Washington, DC, USA.

Kiwanuka, S. and Kintu, J. (2004) Seed security in Uganda: current status, issues and challenges. *Uganda Journal of Agricultural Sciences* 9(1), 17–22.

Larson, D.W. and Mbowa, S. (2004) Strategic marketing problems in the Uganda maize seed industry. *International Food and Agribusiness Management Review* 7, 86–93.

Lele, U. (2009) Global food and agricultural institutions: the cosmology of international development assistance. *Development Policy Review* 27(6), 771–784.

Mpuga, P. (2004) Demand for credit in rural Uganda: who cares for the peasants? Paper presented at the Conference on Growth, Poverty Reduction and Human Development in Africa, Centre for the Study of African Economies, Oxford, UK.

Opio, F. (1999) *Management of Bean Root Rot in Uganda. A Technical Report of a Subproject Funded by African Highland Initiative.* CIAT, Kampala.

Page, S.L.J. (2008) *The Good Seed Initiative: Field Activities with Food Insecure Farmers in South Asia and East Africa.* CABI, Egham.

Rooks, G., Szirmai, A. and Sserwanga, A. (2009) *The Interplay of Human and Social Capital in Entrepreneurship in Developing Countries. The Case of Uganda.* United Nations University, World Institute for Development Economics Research (UNU-WIDER), Research Paper No. 2009/09: 32, Helsinki.

Rubyogo, J.C., Sperling, L., Muthoni, R. and Buruchara, R. (2010) Bean seed delivery for small farmers in sub-Saharan Africa: the power of partnership. *Society and Natural Resources* 23(4), 285–302.

Seibel, H.D. (2003) Centenary Rural Development Bank, Uganda: a flagship of rural bank reform in Africa. *Small Enterprise Development* 14(3), 35–46.

Van Mele, P., Zakaria, A.K.M., Hosne-Ara-Begum, Harun-Ar-Rashid and Magor, N.P. (2007) Videos that strengthen rural women's capability to innovate. *Communication for Development and Social Change* 1(3), 79–99.

11 Madagascar: Coping with Relief Aid and Politics

PAUL VAN MELE, KETAMALALA RANDRIAMILANDY, NOROSEHENO RALISOA AND RAYMOND RABESON

11.1 Introduction

11.1.1 Agriculture

At 400 km from the coast of Africa, Madagascar is the fourth largest island in the world with an area of 587,041 km^2 and about 20 million inhabitants. The capital Antananarivo in the central highlands is linked to the main harbour, Toamasina, by a well-kept road and railway, illustrating the long-term effect of economic interest groups pressuring successive governments to favour food imports.

The climate varies from humid tropical in the east, to hot and dry in the south-west and north-west, arid in the southern regions and temperate in the central highlands, with altitudes ranging from 800 to 1800 metres. Madagascar is often liable to cyclones, flooding and drought, which continue to influence seed demand and supply.

About 70% of the population lives from agriculture. Because of the diverse climates, altitudes and soils, a wide range of food and cash crops are grown. Rice is grown on 1.3 million hectares (about 45% of the cultivated land in 2006), followed by 0.3 million hectares each of maize and cassava. Sweet potato occupies 122,000 hectares. In the arid south, cassava and sweet potato are the main food crops. In the highlands, potato is grown on an estimated 38,000 hectares (MAEP, 2007).

Family plots are generally smaller than a hectare, whereas in the largest rice production areas, such as Lake Alaotra and Marovoay, fields are larger and flatter, allowing mechanization.

11.1.2 Agricultural policy

The government recognizes the important role of improved technologies such as fertilizers and seeds in ensuring food security, the first objective of its Rural Development Policy (2001). The Madagascar Action Plan (2007–2012) seeks to achieve a high-growth economy by ensuring that the country has a diversified and strong private sector driven by local and international investment and trade. Developing input distribution systems is crucial.

Rice markets have been particularly important, since rice is the staple food and producing it is a major source of income and employment. Levels and stability of rice prices have major effects on the welfare of rice farmers (about 60% of the population) and consumers (almost the entire population). Policies to spur the development of rice markets, stabilize prices and provide appropriate price incentives for producers and consumers have varied considerably over time, alternating between government interventionism and market liberalization (Minten *et al.*, 2006).

Major agricultural zones are poorly connected to cities, reducing the competitiveness of local food against imports.

In 2006, the government launched its Green Revolution Plan. Measures such as off-season crops, transplanting and improved rice seeds have reduced poverty and food insecurity prevalent in rural Madagascar (Minten and Barrett, 2008). Paddy production grew by 23% in 2009 due to increased productivity and more land cultivated (MEI, 2010).

Paddy prices have been pretty stable and saw a marked price jump only once (in 2004–2005) when the private sector imports had stopped because of high risks involved in the trade and the government short-run efforts to import were not effective (Dorosh, 2008). As local production increased across the country following the boost of upland rice cultivation and the government's Green Revolution Plan, Madagascar was not affected by the grain crisis in 2008 and paddy prices returned to 450 Ariary ($0.23) per kg. Currently, the government is considering exporting rice so that growers would no longer face low prices at peak harvest times (Philibert Rakotoson, personal communication, 2010).

11.1.3 Evolution of seed sector

Formal seed system. Although as early as 1985 the Malagasy government was committed to getting out of direct agricultural production itself, it continued to guarantee seed production through 20 seed multiplication centres (CMS – Centre de Multiplication de Semence) across the country. Only in 1995, when the seed act was passed, did the state stop subsidizing seed production and begin to privatize the CMS.

Five centres were immediately privatized, but, as companies had only sold seed and never produced it, the initiative flopped. They struggled to survive and could hardly provide seed. Most other centres, as well as most seed growers' associations, stopped producing seed altogether because they lacked credit, once provided by the state. Only the biggest ones – Anosiboribory near Lake Alaotra, Sakay in the mid-west, Anosy (Fianarantsoa) in the eastern highlands, Marofarihy in the south-east, Tsararano (Marovoay) and Mahabo in the west – benefited from donor funds and were able to survive longer. These two last ones are managed by seed growers' associations. CMS Sakay is the only centre specialized in upland crops (MAEP, 2008).

The seed act allows the CMS to also produce foundation seed. The national agricultural research institutes FOFIFA and FIFAMANOR collaborate with international research organizations and should produce only breeder and foundation seed, but, due to the weak organization of the sector, they also produce commercial seed.

Two private foreign companies were established: Semana (a subsidiary of the French Technisem, to produce vegetable seeds for local use and export) and Castells (a Spanish company that started producing certified rice seed around Lake Alaotra in 2007). Also five national enterprises emerged, including Valy Prod Sem and Andri-Ko.

Under the government's Green Revolution Plan, quality seed was considered key to increasing productivity and the Ministry of Agriculture made a special effort to spread quality seed all over the island through its 22 Regional Departments of Rural Development (DRDR – Direction Régionale de Développement Rural). After receiving foundation seed from research, they provide this seed along with training for members of seed producer groups (GPS – Groupement de Paysans Semenciers), whose production is sometimes bought by the Ministry but mostly sold to neighbours.

Most farmers save their own seed, unless they can get a good price for their harvest.

Informal seed system. Most farmers are aware of the advantages of improved varieties, but they know that these do not express their maximum performance without mineral fertilizer. Short of cash, most farmers are unwilling to make that investment. Credit, if available, is rarely considered (MAEP, 2008).

Although most farmers save their own seed or exchange with neighbours, they may buy seed when they get a good price for their harvest. When farmers are unable to save part of their production or when their harvest fails because of natural disasters or crop diseases, many dealers take advantage and sell cheap, low quality seed, with no label or packaging.

The Malagasy Association for Seed Promotion (AMPROSEM – Association Malagasy pour la Promotion des Semences) organizes workshops to promote quality seed and plays a role in seed policy orientation as a member of the National Seed Council (CONASEM – Conseil National des Semences).

11.1.4 Seed law

The seed act 94-038, covering seed regulation, was enacted in 1995, but the implementation decree was issued only in 2006. It aimed to stimulate private sector investment in seed research, production and distribution, import and export, leaving control and regulation to the public sector.

The 1995 law created several seed institutions, such as CONASEM, charged with policy analysis and orientation, seed promotion and diffusion, and the Varietal Admission Committee (CTAC – Comité Technique d'Admission des Variétés au Catalogue), which records cultivated varieties in a national catalogue.

To implement the seed act, a national seed strategy was written in 2007, and, with support from the FAO, legal texts related to the law and a national variety catalogue were drafted. The first edition of the catalogue will be published in 2010 and will include the most popular landraces, local and improved varieties.

11.1.5 Seed certification

Many seed enterprises rely on internal quality control and sell standard seed (which actually is a legal category for vegetable seed only). The Seed Control Service (SOC – Service Officiel de Contrôle des Semences) is currently under the Ministry of Agriculture (although with every new government its name and organizational set-up has changed) and is in charge of inspecting seed fields, laboratory tests, certification and monitoring of seed distribution. SOC's office is in the same building as the National Seed Laboratory, which was established in 1986.

Before 2004, four inspectors were based in the capital, three in the country's main rice production zone (Lake Alaotra) and two in Toliara in the south-west. In 2010, five inspectors were at the head office and 42 in the regions. Certificates are delivered by the head office only. So far, limited amounts of seed have been certified (Table 11.1), but with the legal texts and catalogue being finalized this is likely to change.

So far, SOC has mainly inspected rice, maize, legume and vegetable seeds, free of charge. Inspectors check seed traceability, field isolation, crop rotation and specific and varietal purity and assess the health status of the crop at least twice. Field reports inform growers whether their fields are accepted or not. Inspectors also sample seed growers' postharvest and storage practices, adhering to the standards of the Organisation for Economic Co-operation and Development (OECD).

Table 11.1. Seed certified (tonnes) in Madagascar, 2004–2009. Source: SOC.

<table>
<tr><th></th><th>2004</th><th>2005</th><th>2006</th><th>2007</th><th>2008</th><th>2009</th></tr>
<tr><td>Rice</td><td>1,300</td><td>1,550</td><td>900</td><td>2,045</td><td>1,505</td><td rowspan="8">No seed was certified in 2009 because of the political crisis</td></tr>
<tr><td>OPV maize</td><td>–</td><td>–</td><td>100</td><td>147</td><td>120</td></tr>
<tr><td>Groundnut</td><td>–</td><td>–</td><td>23</td><td>26</td><td>5</td></tr>
<tr><td>Bean</td><td>–</td><td>–</td><td>–</td><td>37</td><td>16</td></tr>
<tr><td>Lima bean</td><td>–</td><td>–</td><td>–</td><td>–</td><td>52</td></tr>
<tr><td>Potato</td><td>–</td><td>–</td><td>–</td><td>48</td><td>–</td></tr>
<tr><td>Vegetables</td><td>3.5</td><td>3.5</td><td>3.5</td><td>3.5</td><td>3</td></tr>
<tr><td>Total</td><td>1,304</td><td>1,554</td><td>1,027</td><td>2,307</td><td>1,701</td></tr>
</table>

Seed producers take samples and send them to SOC, but from 2011 onwards this will be done by the regional seed inspectors to reduce handling and lower costs. The National Seed Laboratory at SOC analyses seed samples according to the rules of ISTA (International Seed Testing Association), although it is not accredited.

11.1.6 Seed demand and production

The seed multiplication centres that are still functional have limited capacity and geographical coverage. They are unable to plan their seed production because seasonal seed demand is not clearly assessed and communicated. Most of the demand has come from the government and relief agencies and responds to natural disasters.

Unions of cooperatives are often poorly organized and reluctant to share information. Various platforms have been established, such as the national rice platform and the potato platform in the Vakinankaratra region, but their role in coordinating seed demand and supply has not yet been explored (MAEP, 2008).

Estimating seed demand from the acreage covered by each crop, certified seed production is low. In 2007 improved rice seed production was 2045 tonnes, which was about 2.5% of the total supply. Certified maize seed production was 147 tonnes, representing about 1% of the total. As seed is multiplied by farmers, a much higher share of production is impacted indirectly.

Rice. Seventy per cent of the national irrigated rice seed is produced by the CMS Anosiboribory near Lake Alaotra. Privatized in 2009, it provides seed even for other regions. Also seed growers' associations supported by DRDR or NGOs produce quality seed and directly supply neighbouring farmers. In 2009 the Ministry imported hybrid rice seed from China, which it has tested and began promoting in 2010. As earlier attempts failed because farm-saved seed of hybrid rice does not give a good crop, it seems puzzling why farmers would not abandon it this time.

Maize. FOFIFA selected varieties adapted to the different agro-ecologies. CMS Sakay and some farmer groups are the main producers of standard maize seed. Fertilis Madagascar imports white maize hybrid seed from Pannar (South Africa).

Potato. Selection and production of breeder and foundation seed are done by FIFAMANOR (a Malagasy–Norwegian cooperation for livestock and crop development). This centre produces standard seed on its own land and also buys from farmers' associations when more seed is needed. Some farmer groups are now producing potato and other crop seed themselves within the Vakinankaratra region.

Beans. FOFIFA is in charge of varietal selection and seed multiplication (CALA near Lake Alaotra). Private enterprises produce commercial bean seed, such as Valy Prod Sem in Miandrivazo (middle west).

11.1.7 Politics, trade and environment

Despite the country's wealth in agricultural and natural resources, in 2005 about two-thirds of the population was living below the national poverty line, with a higher proportion in rural areas (IMF, 2009). Politics has had a major influence on the country's development. Decentralization has been extremely slow, with less than 5% of the overall budget under the responsibility of local governments (IMF, 2009). Roads are good where they serve political interests. Main agricultural zones remain poorly connected to the cities, increasing the costs of local food. Seed enterprises either produce mainly for local customers or are obliged to add a high fee for transporting seed to other areas.

Agricultural intensification linked to increased population pressure in the central highlands has caused soil degradation. Deforestation has led to erosion with downstream consequences. Rivers, which used to add fertile silt during the annual floods, have started to deposit sand on the agricultural land, drastically reducing soil fertility. As mineral fertilizers are expensive (compared with seed and grain), seed producers still rely mainly on the natural restoration of soil fertility and will increasingly be challenged to stay in the seed business. Also, irrigation channels such as those in the main rice producing region of Lake Alaotra, need increased maintenance due to the sand deposits.

After deforestation rivers carry more sand and irrigation canals become clogged. Floods no longer deposit silt, which once restored natural soil fertility.

Over the past decade donors have promoted farmers' associations and cooperatives, but most have become dysfunctional shortly after the projects ended. Many businesses in Madagascar, whether for seed or other enterprises, are family-based.

The section below presents three family-based seed enterprises: Valy Prod Sem produces bean seed for local and export markets; Andri-Ko grows cereal seed, mainly irrigated rice; and a farmers' association near Antsirabe grows sweet potato vines, among other crops. We also describe various government institutes that are privatized (SCAA), completely self-financed (CMS – Sakay) or partially financed (FIFAMANOR). The final case describes a potato seed growers' cooperative.

11.2 Valy Prod Sem

11.2.1 History

During the big privatization in the late 1980s, Mr Mbosa Rabenasolo, being a senior government official, received the rice seed multiplication centre in Anosy-Avaratra, which was converted into a cooperative. However, faced with frequent floods and land tenure problems with neighbouring farmers, he abandoned it to start his own enterprise with his brother-in-law, Mr Liva Nirina Rasata, in 1990. They christened the company Valy-Agri, *valy* meaning brother-in-law. (In Malagasy culture women are respected and the man who marries a sister is treated courteously.)

Rather than rice, the new company began trading in bean seed for export. They started collecting dried beans near Miandrivazo in the hot mid-west and shipped two containers of 22 tonnes to France and the United Kingdom. Having been director of the seed and plant material service (SMV – Service des Semences et Matériel Végétal) at the Ministry of Agriculture, Mr Mbosa had all the necessary contacts. Moreover, he had a brother in France, who facilitated links with importers.

However, soon the French importers started to complain about varietal mixtures. Mr Mbosa realized that the bean growers did not use quality seed. Being a seed technologist by training, he started collecting seed from farmers in the region, purified and conditioned them and handed out about 25 tonnes of bean seed to farmers per year to ensure the delivery of a uniform product, required by the European market. Farmers sold their crop to Valy-Agri and returned one and a half times the amount of seed received.

The brother-in-law, Mr Liva, who was trained as an electro-mechanic, learned all about seed production and started training farmers. However, in 1996 (when the grain market price exceeded the price Valy-Agri had fixed for seed at the start of the season) most farmers failed to return the seed as per the contract, so Valy-Agri stopped working with them, except for a few committed farmers.

After contracts were not respected, Mr Liva (right) continued working with a handful of faithful bean seed growers, such as Daniel Rakotoniaina (left).

From the mid-1990s the French seed company Technisem contracted Valy-Agri to produce seed of French bean. Technisem already grew vegetable seed (mainly for the European market) in Antsirabe, at about 1500 m altitude, but needed more land to grow bean seed. The collaboration with Valy-Agri benefited both parties and was sustained for nearly a decade. Mr Liva began producing seed on his own land (about 7 hectares), but later lost this following the cyclone of 2004. Moreover, during the civil war in Côte d'Ivoire, Technisem stopped the collaboration. Ever since, Valy-Agri has produced seed for the domestic market with some faithful seed growers.

Apart from the French bean Nain Bobby, which has no string and a very good taste, whether cooked or eaten raw, the company also grows Royal Special. The bean Lingot Blanc has been a hit from the beginning. More recently the bean Marbré Rouge was added. The company has continued selling dried beans (Table 11.2). Until 2006, it bought from Andri-Ko maize, rice and groundnut seed, for which it had found a market.

Table 11.2. Seed produced (tonnes), Valy Prod Sem.

	2004	2005	2006	2007	2008	2009
French bean	9	2	–	1	–	1
Beans	11	7	10	25	37	35
Groundnut	–	–	8	2	–	3
Soybean	–	–	–	3	5	–
Total	20	9	18	31	42	39
Dried beans for local consumption (tonnes)						
	20	30	30	30	30	30

Seed is quality declared, but not certified as fields were not inspected.

Over the years Mr Mbosa established four enterprises: the seed enterprise Valy-Agri, a consultancy bureau, a communications company and a music business. Having all enterprises grouped made it impossible to know which one was truly profitable and which one not, and became a nightmare for the bookkeeper, so from 2009 all enterprises began operating independently. Mr Livo became director of the seed enterprise Valy Prod Sem.

11.2.2 Structure

Management. Mr Liva is director and the only member of staff, advised by Mr Mbosa, plus 50 to 100 temporary workers, as needed. Although the local market for bean seed has grown, Mr Liva's main objective is to move back into the export business.

Outgrowers. Although it started working with many outgrowers, in 2010 Valy Prod Sem worked with only one farmer group and two individual seed producers, covering about 30 hectares in three villages (Box 11.1).

Land. Seed is produced on farmers' land only. Daniel Rakotoniaina and his brothers have about 10 hectares of irrigated land and 28 hectares of land along the Mahajilo river, in Masiakampy village, a few kilometres from the main road to Mioandrivaso, which was tarred only in 2008. The land is flooded every year by the river, which deposits alluvial soil (locally called *baiboho*) and adds natural soil fertility. However, bush fires in the region have stripped the land of trees, resulting in serious problems with erosion. When the river deposited sand on top of the fertile land, yields dropped from 1.8 tonne to 500 kg per hectare.

Infrastructure and equipment. In 1992, the company had two tractors and seeders, which they loaned to others. They sold them about 10 years later as their own seed farmers were reluctant to use the heavy machinery as it destroyed the structure of their soil. The company used to have its own truck, but given the poor conditions of the roads it broke down every year, so they sold it. The limited capacity of the enterprise most likely made it more profitable to outsource this service. Valy Prod Sem currently pays a 25-tonne truck to collect and haul the seed to the capital, about 240 km away, where it can clean and store 40 tonnes.

Quality control. All farmers have been trained in seed production. The production zone has so far been free of diseases and no fertilizers or pesticides are used in the fields, which are manually weeded.

Until recently the company only sent samples to the seed laboratory without having its fields inspected. Hence, seed is not certified. From 2010 onwards SOC will only analyse samples of inspected fields.

For the Lingot Blanc bean, producers keep their own seed every year. Valy Prod Sem gave them a cleaner-grader and a ventilator to clean their seed themselves. However, for the other varieties it supplies new seed every year. Valy Prod Sem treats its seed with K-Otrine (deltamethrin) against storage pests before putting it in 25 kg bags.

Box 11.1 A mason turns to seed.

Daniel Rakotoniaina studied to be a mason, but he always had a passion for agriculture, so his lawyer uncle gave him land in Masiakampy.

Having bought a bag of bean seed in the capital Tana in 1986, Daniel multiplied it, but noticed that it did not give good results. So in 1990 he decided to start all over again. He went for training from SMV and bought 100 kg of Lingot Blanc bean seed from Bera-Agro, a French company.

One lady in his family worked for the seed multiplication centre in Sakay. She asked Daniel to monitor the on-farm varietal performance tests for maize and groundnut and taught him a few more things. At the end, Daniel kept some of the seed, which he multiplied for 5 years. But as he had no customers he gave up on it. His commitment and eagerness to produce quality seed came to Mr Liva's attention, so since 1996 he and his brothers have produced bean seed for Valy-Agri. As his reputation grew, Daniel took up maize seed production again in 2007, adding groundnut in 2009. His eldest son is at the university. 'He studies agriculture to do even better than me,' says Daniel, unable to hide his pride.

11.2.3 Cash flow

For 2 years Mr Liva took a bank loan of 25 million Ariary ($12,500 at current rate, but worth about $20,000 at the time) to collect seed for export. Once the importers had paid, generally within a month after they had received the container, he repaid the loan. Mr Liva has only used his and his wife's savings to invest in the company's assets.

11.2.4 Marketing

Having four linked businesses may have been hard on the bookkeeper, but it had its advantages. The Valy communication business distributed diaries (or appointment books) with a different agricultural theme each year. The diary contained adverts for the Valy seed business and targeted literate people. This made the company known to NGOs and projects, who have remained the principal clients ever since (Table 11.3), although the crisis in 2009 has reduced the number of NGOs working in the country. In 2007 Agro-Action Allemande still bought 100 tonnes for seed relief in the south and south-western regions.

Table 11.3. Clients of Valy Prod Sem.

	1992	1999	2004	2009	2015 (predicted)
Projects and NGOs	1	1	1	1	–
CMS Nansan	–	3	3	2	–
Export	–	2	2	–	–

Ranking assessment by senior management of seed enterprise, 1 being the most important. For 2015 no projection was made as management felt the future of its business was entirely dependent on the political and economic situation.

Valy Prod Sem also has a sales depot at the seed multiplication centre (CMS – Centre de Multiplication de Semence) in Nanisana (in Antanarivo), which sells seed directly to farmers. Mr Liva did not want to make any future projections, as 'politics determine the economy of our country'.

11.3 Seed Multiplication Centre (CMS) – Sakay

11.3.1 History

The Seed Multiplication Centre (CMS) in Sakay, about 155 km west of the capital along the national highway number one, was the last one established among the 20 centres spread across the country. The CMS started as a project to promote maize in the mid-western region in 1990.

Five years later, after its first evaluation, the centre received a national mandate to produce maize seed for the entire country. In 1999, the maize project came to an end and the centre was obliged to finance itself. With the exception of the director, Mr Rafalimanana (Rafaly), who continued to receive his salary from the government, all staff had to be paid from the centre's revenues.

The centre grows seed of maize (meva, IRAT 200, volasoa and NTS 10), groundnut, bambara nut (*Voandzeia subterranea*) and upland rice (B-22, 3737, 3728, FOF 154, Nerica 2 and Nerica 4) on its own land (Table 11.4). For seed of legumes and irrigated rice (X-265, X-243, 2067, MGK, Tsemaka and Sebota), CMS started collaborating with seed producer groups in 1992. Being mainly driven by project support, production with these groups was interrupted at times, but has continued ever since 2002. Soybean seed was produced once only in 2005. As some businessmen have recently established export markets for beans (Lingot Blanc) demand for seed is going up rapidly.

11.3.2 Structure

Management. It has been hard to fully self-finance the centre for the past 10 years, but the director Mr Rafaly remains positive and hopeful that things will change for

Table 11.4. Seed produced (tonnes), CMS – Sakay. Source: SOC and CMS Sakay.

	2004	2005	2006	2007	2008	2009
Maize	16 (1)[1]	80 (4)	60 (3)	50 (3)	70 (3)	85 (3)
Rice (irrigated)	20 (4)	50 (6)	20 (4)	60 (4)	80 (4)	26 (1)
Rice (upland)	25 (2)	45 (7)	18 (4)	42 (4)	40 (4)	105 (2)
Groundnut	2 (1)	2 (1)	2 (1)	2 (1)	2 (1)	2 (1)
Other legumes	2 (2)	10 (3)	6 (2)	4 (2)	5 (2)	6 (2)
Total	65 (10)	187 (21)	106 (14)	158	197	224 (9)

[1]The number of varieties is given in parentheses.

the better. Until 1997, seed prices followed the same trend as those of petrol and fertilizer: 1 kg of seed cost as much as 1 litre of petrol or 1 kg of fertilizer. In 2010, a litre of petrol cost 2500 Ariary ($1.25) and a kg of fertilizer 2000 Ariary ($1), whereas certified rice seed cost only 1500 Ariary ($0.75). To be sustainable, seed prices ought to go up. However, seed companies decide on seed pricing, but fear that farmers do not want to pay more for seed, particularly since the government keeps the price for paddy as low as 400 Ariary ($0.20) per kg. Mr Rafaly believes that if seed prices had followed the same trend as fuel and fertilizer prices he would have been able to renew the centre's equipment, which is deteriorating after 20 years of service.

Apart from the director, there are 12 permanent staff and 60–80 temporaries during peak periods. Throughout the year about 8000 person-days of day labourers are paid.

Outgrowers. There are no outgrowers in the strict sense, but the centre supervises four farmer seed producer groups located in its direct vicinity. These groups produce irrigated rice seed, as well as seed of legumes such as bambara nut, groundnut and beans. They are given foundation seed that the centre sources from FOFIFA and are offered a good price to ensure that they sell back the seed they grow. Seed from these producer groups is sold through the same outlets as seed produced by the CMS.

Land. The centre has about 200 hectares of land. Since 1992, 10 hectares in Sakay have been used as showcases and demonstration plots along the national highway number one to show the new varieties along with new cultivation practices. Ten years later the Ministry of Agriculture copied this promotion strategy. The bulk of fields for seed production are in Imehy and Diavolana, but under current conditions the CMS has only been able to grow seed on 70 hectares.

Infrastructure and equipment. These are 20 years old and starting to deteriorate. The herbicide spray tank is crucial, since upland seed crops require timely weed control. The centre has its own generator and so is self-reliant in electricity, as in various other cases (e.g. Sections 10.2.2 and 11.3.2). The seed processing plant has a capacity of 10 tonnes per day. Seed growers who want to make use of it pay 120 Ariary ($0.06) per kg seed.

Links and partnerships. The centre obtains its rice foundation seed from FOFIFA, but the CMS maintains its own maize foundation seed, for which Mr Rafaly was trained. FOFIFA only provides maize foundation seed when there are project funds, so the CMS in Sakay cannot rely on them for maize.

The CMS is a member of AMPROSEM, but feels too little is done and that workers in the seed sector need to discuss more openly the problems faced by the sector. It believes that the lobbying work of the national rice platform, aimed at keeping down rice consumer prices, is counterproductive for the seed industry.

Quality control. Although the centre only produces seed for upland crops, it directly supervises the four farmer seed producer groups in its vicinity that also produce seed for irrigated rice. They receive foundation seed and are required to adhere to rigorous standards: line transplanting, as well as timely roguing, weeding and harvesting and proper drying. Farmer seed producer groups are motivated to produce quality seed by giving them a decent price, 800 Ariary ($0.40) per kg, double the price of paddy. Processing and transport further add to the final seed price.

Quality control is done by the centre itself and follows the procedures for certified seed production, including the use of pure foundation seed, respecting isolation distances and proper roguing, weeding, drying and storage on wooden slats. Samples are only sent to SOC when clients require this. Even then, fields are not checked by official inspectors.

Mr Rafaly stresses that retailers need training to ensure seed quality is maintained during storage, whereas farmers need to be made aware of how to recognize certified seed to prevent retailers refilling seed bags with standard paddy.

11.3.3 Cash flow

'We only give credit to those retailers that we know. They have to repay the first half after a month, and the second tranche after 2 months,' explains Mr Rafaly. 'To those we do not know, we do not give credit, but we give them a slightly lower price, so they feel motivated to buy from us.'

Given the political crisis and high interest rates, the centre has never considered taking a loan to upgrade its infrastructure and equipment. Since its herbicide sprayer broke down in 2009, it has bought spraying services from others. In return, CMS lends them its tractor and seeder.

11.3.4 Marketing

New varieties are advertised via agricultural fairs and occasionally also via the Radio National Malagasy (RNM) and TV clips. Since its early days CMS has had a shop in the capital, near the offices of the seed certification service. This shop became more important when the centre became financially self-sustaining in 1999 (Table 11.5).

More recently, orders are increasingly placed directly with the centre in Sakay by means of the mobile telephone or email. Projects and NGOs have remained important clients and often order seed at a negotiated price that includes the cost of transporting the seed to the desired destination. Governments only buy seed after disasters and during elections to win votes. However, they often only pay after 2 years, which ties up the centre's operating capital (as is the case in Nigeria, Section 4.1.3).

The retailer network is still poorly developed, although retailers are important because of their proximity to farmers. Most retailer shops with whom CMS Sakay works are run by the farmers' association of the mid-western region, with whom it has developed a trust relationship over the last 20 years. In 2009, nine of their shops sold about 12–15 tonnes of maize and rice seed. Apart from these, CMS sold about 8 tonnes of seed through two agro-dealers in the capital. Retailers are important for maize and rice, but are unwilling to stock seed of legumes.

Table 11.5. Clients of CMS – Sakay.

	1990	1999	2005	2009	2015 (predicted)
Government	–	4	4	1	4
Projects and NGOs	1	2	2	2	1
CMS commercial shop	2	1	1	3	2
Individual farmers	3	3	3	3	3
Retailers in the region	–	3	3	4	5
Commercial maize growers (> 3000 ha)	–	–	–	4	–
Retailers in other regions	–	4	4	5	6

Ranking assessment by senior management of seed enterprise, 1 being the most important.

A major shift in marketing has been the introduction of small seed bags of 1, 2 or 3 kg. Although in Uganda this was mainly a response to overcome counterfeit seed sales and to make seed more accessible to smallholder farmers (Section 10.2.4), in Madagascar this mainly followed the adoption by farmers of new rice cropping technologies, such as transplanting, which requires less seed than broadcasting. With the system of rice intensification (SRI), farmers require only 6–7 kg of seed per hectare, so they want to buy even smaller quantities. Although pre-printed plastic bags could be ordered from a packaging company, the CMS buys rolls of plastic and heat seals its own bags to reduce costs. Apart from the small plastic packs, upland rice is also available in 10, 25 and 50 kg bags, whereas irrigated rice and maize seed is no longer sold in 50 kg bags.

About 50% of the clients are repeat customers who buy seed every year, but this proportion is generally higher for upland rice farmers than for irrigated rice farmers. In the warmer mid-west, upland rice is harvested during the hunger period, when the stock of irrigated rice is low. Upland rice farmers obtain higher prices and therefore sell most of their harvest and buy seed every year.

Projects are the main buyers of irrigated rice seed; sales dropped in 2009 when the political crisis meant that many donors stopped funding activities in Madagascar. Half of the rice seed produced had to be sold as paddy. For the CMS Sakay, seed sales of other crops were not affected by the crisis. Although it has been operational for 20 years and has been fully self-financing for the last 10, Mr Ravaly believes that the long-term sustainability of the seed centre is in danger if paddy prices are kept low.

11.4 SCAA

11.4.1 History

When Mr Andry Rakotovao returned from France in 1985, where he studied mathematics, he started his own little business by collecting rice from farmers. Son of an agricultural engineer who worked for the research institute in the capital, he often spent his school holidays at his uncle's place near Lake Alaotra. His uncle owned a rice mill and a few trucks to collect rice from farmers. As a kid Andry often helped his uncle and dreamed of becoming just like him one day. Driven by a sense of self-determination Mr Andry bought his first small truck in 1987 and 3 years later he bought a rice mill.

In 1998 he embarked on rice production and established the Société Commerciale et Agricole d'Ambohimangakely (SCAA Ltd) to formalize his enterprise. Mr Andry never took a loan from a bank to invest in his business, but always relied on his own savings.

'Changing the mentality of civil servants after privatizing the state seed multiplication centres was a real challenge,' says Mr Andry, the director of SCAA.

Over the years his transport company expanded to about 40 trucks. Mr Andry also established his own media company, as well as his own import–export company (only for rice), and became president of the national rice platform in 2005.

In 2009 Mr Andry started managing the run-down seed multiplication centre (CMS – Centre de Multiplication de Semence) in Anosiboribory, near Lake Alaotra, at an altitude of 700 m. He took over 100 of its permanent staff, along with 1000 temporaries. Some staff were made redundant, mainly to get rid of the civil servant mentality, and replaced by more dynamic and motivated staff. Strict budget plans were introduced to improve the financial status of the centre and Mr Andry brought in new suppliers for fuel and fertilizer. Within a year Mr Andry invested about $400,000 to improve the Centre's infrastructure and equipment. Rice yields doubled from 2 to about 4 tonnes per hectare. In 2009 he produced 400 tonnes of certified rice seed, but he could sell only 100 tonnes. The rest he milled and sold as white rice.

A seasoned businessman, but a newcomer to the seed sector, Mr Andry learned two major lessons from his first year's experience. Apart from productivity, he will improve the quality of production, so that, in case he cannot sell all as seed, at least he will be able to sell it as luxury rice, sold at twice the price of ordinary white rice. He also plans to advertise the company's seed more aggressively and increase its distribution network from Antananarivo, Tamatave and Analamanga regions to all the rice-producing regions of the country. Since he has already established his own transport and media companies, he believes these will help him to achieve this goal.

Mr Andry has maintained close relations with the government and produces seed of crops and varieties requested by the Ministry of Agriculture. In 2011, he intends to multiply cassava cuttings and seed of maize, upland rice and soybean.

11.4.2 Structure

Management. The seed multiplication centre underwent a major change in management, shifting from a public servant style to an enterprise with rigorous budget plans and strict quality control (both technical and financial).

Managing over 550 hectares of government-owned land to grow rice seed requires professionalism, flexibility and firmness.

Staff. Apart from the director general, a director of operations and a director of administration and finance, there is a person in charge of the field operations and one in charge of the seed plant. The young guards are alert and wear fresh beige and red uniforms. The facelift of the seed processing plant only partially reflects the on-going, fundamental shift in attitude.

Outgrowers. SCAA does not work with outgrowers, but in 2009 it established a contract between the Direction Régionale de Développement Rural (DRDR) and various seed producer groups, who were given 100 hectares to produce seed. SCAA prepared the land for them while DRDR trained and supervised them. Although there are no reciprocal obligations, most of the seed is processed by SCAA for a fee, negotiated at the time of harvest.

Land. The centre has 559 hectares of irrigated land. With the lease contract, Mr Andry manages all of it, apart from 50 hectares that remain under government control, mainly to maintain the public status of the CMS. However, Mr Andry took over the management of the whole infrastructure, including the warehouse, so the government now has to negotiate the terms and conditions with SCAA to make use of it. In 2009, the government produced rice seed on only 25 hectares, as it had to learn how to work with the private sector.

Infrastructure and equipment. The warehouse can store up to 1000 tonnes of seed. SCAA took over all the old machinery, including the combine harvester, and bought three new Indian tractors, a generator and new computers for the office. The Japanese seed processing unit dates from the early 1990s and will soon be replaced by a smaller, modern unit. The seed plant and offices of SCAA are not linked to the power grid, but are powered by a generator.

Quality control. All fields and seed are officially inspected in order to obtain certified seed. From 2009 SCAA started hiring Mr Rakotobemalala, one of the five inspectors working in the region at DRDR, who is close to retirement, as adviser. It had an immediate effect on the total number of fields approved for seed production.

11.4.3 Cash flow

Mr Andry borrows money from various banks for operations, but has never taken a loan for capital investment. For this, he relies on his own savings. However, he intends to take a loan in order to renew the seed processing equipment.

'Combining seed with commercial rice production is a good strategy, as it helps to free up money more quickly,' explains Mr Andry. The milled rice can be sold immediately after harvest, improving cash flow, while seed is stored for sale during the next season.

11.4.4 Marketing

SCAA produces seed of multiple rice varieties, although the bulk is the local variety *makalioka* (MK 34, which was released in 1934) and the more recent improved variety 4012 (*tsemaka*, which is a cross between MK 34 and 2787, a variety that tillers profusely). MK 34 has remained the most popular variety in the region. The long panicles have many ramifications, resulting in good yield. The slender grains swell well after cooking and have a well-appreciated taste. As consumers in the capital mainly ask for these varieties, the collectors mainly search for this rice when they drive their trucks over the 250 km long and difficult road, half of which is winding through the mountains, the other part being a degraded sandy road.

In 2009, SCAA sold 30 tonnes to the Ministry of Agriculture and 40 tonnes to farmers. The remaining 30 tonnes remained unsold and were milled and sold as white rice.

SCAA set up its retailer network around Lake Alaotra by relying on existing shops, either grocery shops, wholesalers or agro-dealer shops. Rice seed bags of 25 and 50 kg are clearly labelled and contain variety information. Posters with photos of the different rice varieties were produced and given to the shops willing to sell SCAA seed, along with flyers and brochures explaining certified seed and its benefits.

Understanding the importance of media, Mr Andry had various radio programmes made and broadcast by the local stations Amomix and Relax. He intends to strengthen further the seed marketing in the future.

11.5 Andri-Ko

11.5.1 History

During his service at the Ministry of Agriculture Mr William Andriamasinoro worked as a technician alongside rice researchers and for over a decade supervised activities of the country's main seed multiplication centre at Anosiboribory, near Lake Alaotra.

He benefited from training courses and exchange visits to Morocco, Kenya and Australia. After working for 35 years, Mr William and his wife, who also worked as an agricultural technician, decided to take early retirement and establish a family enterprise. While his wife looks after the fruit tree nursery and flower production, Mr William cashed in on his skills and relations and set up a seed enterprise.

Lake Alaotra is Madagascar's primary rice production zone, with about 160,000 hectares of rice fields, half of which are irrigated by the river Maningory (and on which two rice crops are possible), so the prospect of producing and selling rice seed looked promising. In 2002 Mr William registered Andri-Ko as a cooperative, which in fact comprises 11 family members. Its premises are in Ambatondrazaka. A year later Andri-Ko produced 20 tonnes of seed of some popular upland rice varieties (2366, IRAT 134 and IRAT 112) that had been released by FOFIFA.

Releasing a few new varieties every 3 years is part of Andri-Ko's strategy to stay in business. It also continues producing seed of the local variety *makalioka*, which performs well without mineral fertilizer.

While increasing the seed production every year, grain prices kept on going down so farmers became more reluctant to renew their seed. Mr William quickly realized that he needed new varieties to stay in business. From the local varieties he began to select the most appreciated and performing ones that were not in FOFIFA's seed catalogue and started to clean and multiply them. By 2010, Mr William had established a collection of about 60 rice varieties, of which he intends to release only a few varieties every 3 years.

In 2007 Mr William followed the government's move to boost upland rice production and started producing Nerica 4 seed. Although it performed well and matured in only 90 days, he abandoned it because farmers who had first adopted the variety had lost interest after they learned that its taste was not fully appreciated and that it was hard to thresh. The people who threshed the Nerica 4 complained that its dust was very itchy. 'As far as I remember, rice varieties that are hard to thresh have never been adopted by farmers in Lake Alaotra,' says Mr William.

By 2006 he also started producing maize seed and in 2009 he sold 1 tonne of seed of legume crops (Table 11.6). These included vetch (*Vicia ervilia*, an ancient grain legume crop of the Mediterranean region), lablab bean (*Dolichos lablab*), cowpea (*Vigna unguiculata*) and some pois du Cap (*Phaseolus lunatus*, a white bean of Malagasy origin that is mainly grown in the south-western region of the country). All of these were basically grown as cover crops in between the maize and sorghum.

At one stage Mr William sold rice foundation seed to the FAO, who wanted to supply their newly trained farmer seed producers. It took a lot of effort to produce, and they bought only once. In future, he will refrain from producing and selling foundation seed.

Table 11.6. Certified seed produced (tonnes), Andri-Ko. Source: SOC and Andri-Ko.

	2005	2006	2007	2008	2009
Rice (irrigated)	50 (4)[1]	60 (6)	137 (11)	155 (18)	70 (9)
Rice (upland)	–	15 (3)	15 (3)	30 (3)	40 (3)
Maize OPV	–	25 (4)	25 (1)	–	50 (1)
Sorghum	–	–	–	–	20 (1)
Groundnut	–	18 (2)	–	–	–
Beans	–	–	–	–	1(1)
Total	50 (4)	118 (15)	177 (15)	185 (21)	181 (15)

[1]The number of varieties is given in parentheses.

Although most technicians in Madagascar continue to aim at introducing short cycle rice varieties, Mr William does not share that vision. Over 70% of the farmers around Lake Alaotra continue to grow the local variety *makalioka* (MK 34). It is photosensitive, has a cycle of 180 days and yields 3.5–4 tonnes per hectare without mineral fertilizer.

During his time at CALA (Complexe Agricole du Lac Alaotra, a regional station of FOFIFA), Mr William acquired a sense of the need to develop new varieties. However, rather than replacing the popular varieties, such as MK 34, his main innovation strategy is to identify and promote new ones that are suitable for niche markets, such as for rice fields with poor water management (about 50,000 hectares around Lake Alaotra), or for newly exploited fields. With increased pressure to survive, fisherfolk have started growing rice on about 2000 hectares at the border of the lake (transplanting as the water in the lake recedes after the rainy season). Upland rice varieties adapted to a wide range of conditions and that perform equally well under irrigation, such as the Sebota varieties developed by French breeders in Brazil, have been added to Andri-Ko's varietal portfolio and will be the next ones the cooperative starts promoting.

The street part of the office has been refurbished as a mobile phone shop (named Andri-Com), showing the uncertainty of running a seed business in a difficult political and economic climate, as well as the adaptability of the overall family enterprise.

11.5.2 Structure

Management. The president is currently backed by his father, an accountant, a bookkeeper and a secretary. There is one technician in charge of rain-fed crops and one for irrigated rice. Three staff members coordinate the work of the 100 labourers that are mobilized during peak periods, such as transplanting and harvesting. One person is in charge of the warehouses, along with a guard.

Infrastructure and equipment. Andri-Ko has two warehouses, each with a capacity of 50 tonnes. One is at the cooperative's headquarters, in the centre of town, while the other one is near the maize fields (maize is stored on the cob, until orders are placed). The cooperative owns a few rotary weeders and two Kubota moto-cultivators that are used for preparing the land, threshing the rice and transporting sand, firewood or people during the off season. They are also used to power the maize sheller. Apart from this, Andri-Ko has two winnowers, which at times farmer seed producer groups and other seed companies rent.

Land. Andri-Ko has 50 hectares, both upland and irrigated. It has some other family land, but this is not yet used for seed production. It is kept under permanent ground cover, using leguminous cover crops to protect the slopes against erosion.

Outgrowers. Loose agreements exist with five outgrowers who can grow at least 5 hectares of irrigated rice seed and who have their own storehouse. Andri-Ko offers 200 Ariary ($0.10) per kg on top of the market price for paddy, but there are no reciprocal obligations. This flexibility suits both parties as there are too many uncertainties.

Links and partnerships. Andri-Ko is a member of the Malagasy Seed Producers Association AMPROSEM. Although Mr William has good relations with research, he exploits his own collection and varieties recently introduced from Brazil. Likewise, he has good relations with DRDR, but prefers to do as much as possible of the quality control himself, to keep costs down.

Seed inspection and certification require training and firm follow-up.

Quality control. The five outgrowers have been trained by Mr William, but are requested to organize and pay for their own seed certification. Andri-Ko only buys seed that has been certified. For seed production on their own land, Mr William and his son rigorously control quality in the field, backed by two supervisors. Andri-Ko pays for official certification by the seed inspector and performs its own seed germination tests before sending samples to the central laboratory at SOC in Tana.

Maize seed is always treated with either deltamethrin or sumithion against storage pests. When the stock is not sold within a few months, an additional fumigation with phosphine may be done. Rice seed is only treated if requested by the client.

11.5.3 Cash flow

Mr William never took a loan. According to him 'micro-credit institutions are often portrayed as helpers of the poor, but many farmers have gone bankrupt because of them. Interest rates are just far too high. It is authorized theft.'

Andri-Ko sells rice seed at 1400 Ariary ($0.70) per kg, which gives it a narrow profit margin. It sells upon cash payment and does not provide credit for its clients.

If needed, Andri-Ko mobilizes capital from its 11 (family) members. In 2009 the FAO contracted Andri-Ko to deliver 20 tonnes of sorghum seed (IRAT 204) to the southern part of the country; Mr William's family had to mobilize 12 million Ariary ($6000), or 600 Ariary ($0.30) for each kilogram of sorghum seed.

11.5.4 Marketing

Mr William wanted quality seed to be available to the farmers, so from the very beginning he established links with 14 agro-dealer shops in all major centres around the lake. As most shopkeepers cheated or disappeared with the seed stocks or agrochemicals and even with the cash, only two remained operational after 2008. Their importance as markets for Andri-Ko has drastically fallen (Table 11.7).

Table 11.7. Clients of Andri-Ko.

	2003	2009	2015 (predicted)
International relief agencies	–	1	–
Individual farmers	2	2	–
Agro-dealers	1	3	–

Ranking assessment by senior management of seed enterprise, 1 being the most important. For 2015 no projection was made as management felt the future of its business was entirely dependent on the political and economic situation.

Individual farmers have kept on visiting the cooperative's two warehouses to purchase seed. Although seed is generally sold in 50 kg bags, Mr William admits that very few rice farmers buy such large quantities nowadays, partly because the landholdings have reduced in size and partly because broadcasting has been replaced by transplanting (widely promoted by government and NGOs and which requires about five times less seed). At the cooperative, bags are readily opened to sell farmers the quantities they desire, mostly 5–10 kg.

In 2006 the NGOs CARE International and the Catholic Relief Services (CRS) placed large seed orders to supply the cyclone-affected areas, while in 2007 the FAO also became a client.

Mr William targets selling seed of new varieties that he wants to release every 3 years. For this he is working closely with Dr Seguy, one of the three researchers (Seguy, Bouzinac and Taillebois) who developed the Sebota varieties in Brazil.

He obtained a contract from the BV Lac project (Projet de Protection des Bassins Versants du Lac Alaotra) to produce 500 kg of seed in small packs of 250 g. The project had already tried out such test kits for sorghum and cover crops and now wanted to try it for rice. Mr William believes that, when farmers have tried it on 100 square metres by following the advice given on the accompanying technical leaflet, they will come back to him the year after to buy larger quantities of seed of these new varieties.

Although Mr William did not want to anticipate how the future of their family seed business would turn out and who would be the main clients in 2015, he remains hopeful; he even handed over the presidency of the cooperative to his son Laingo.

11.6 FIFAMANOR

11.6.1 History

FIFAMANOR is a centre for rural development and applied research created in 1972 from a convention between the Malagasy and Norwegian governments. Evolved from a project, FIFAMANOR became a public utility company with financial and administrative autonomy (EPIC – établissement public à caractère industriel et commercial) in 1992. The Malagasy government pays the salaries of all staff. The Norwegian government contributed to the centre's core funds, based on 5-year plans, but suspended its support in 2009, following the political crisis in Madagascar.

Most of the investments, including the laboratories, screen houses, warehouses and offices, are from Norway. The French Development Agency (AFD - Agence Française de Développement) donated a combine harvester in 1997. In 2007, the PSDR project (Projet de Soutien au Développement Rural) built another potato warehouse.

The centre coordinates research, production, diffusion and marketing of potato, wheat and other cereals, sweet potato, forage and milk production. FIFAMANOR has evaluated potato, wheat and forage varieties in different agro-ecologies for over 20 years.

Seed production has remained relatively stable, although various factors affected seed production in 2009 (Table 11.8). Bacterial wilt became such a problem that new fields had to be sought. During the political crisis, grain prices for maize and soybean became higher than FIFAMANOR's price for maize and soybean seed, so the seed produced on-station was used as animal feed.

Seed of four upland rice varieties are grown on-station, namely FOFIFA 159, FOFIFA 161, FOFIFA 172 and *chom rong dhan* (a Nepali variety). However, as upland rice in the central highlands is harvested at the same time as irrigated rice, it does not command a higher price (as is the case in the mid-western region (Section 11.3.4)) and so relatively few farmers are interested in buying seed. So, although

Table 11.8. Seed produced (tonnes), FIFAMANOR. Source: FIFAMANOR.

	2005	2006	2007	2008	2009
Potato foundation seed	66 (7)[1]	22 (5)	45 (5)	26 (7)	11 (4)
Potato	181 (2)	147 (2)	114 (2)	197 (3)	63 (2)
Soybean	5 (3)	22 (6)	12 (2)	20 (2)	–
Maize OPV	16 (2)	34 (2)	7 (2)	5 (3)	–
Wheat and triticale	6 (2)	10 (2)	8 (4)	30 (4)	3 (4)
Rice (upland)	1 (4)	6 (4)	7 (2)	18 (3)	13 (2)
Total	275 (21)	241 (21)	193 (17)	296 (22)	90 (12)

[1]The number of varieties is given in parentheses. Data do not include seed of forage species.

upland rice production increased under the past government's agricultural policy, only part of the seed produced could be sold.

11.6.2 Structure

Management and staff. There are five departments: for research, production, extension, marketing, and administration and finance. Of the 139 staff, 13 are public servants.

The research department is in charge of breeder and foundation seed production of potato, sweet potato, wheat and triticale. For potato seed multiplication, FIFAMANOR uses its high altitude station (2400 m). For sweet potato, the case is different because it is difficult to store vines. The research department multiplies vines in farmers' plots in Mandoto, Betafo and Manandona, and in some parts of Vakinankaratra, where frost is very scarce (Box 11.2). The extension department has nurseries and demonstration plots in different parts of Vakinankaratra. Farmers take care of the plots and sell the vines.

The extension department is the largest, with 47 staff. These include a network of 29 extension agents who live in the rural areas and who maintain close relations with farmer seed producer groups. Agents meet at the centre once each month. The marketing department has a stock keeper and a person in charge of selling seed, milk and dairy products.

Outgrowers. For years FIFAMANOR worked systematically with farmer seed producer groups, but when market prices went up farmers sold their seed on the market rather than to them. Hence, it stopped using production contracts for upland rice in 2000 and for potato in 2006. Seed growers now buy foundation seed from FIFAMANOR and can sell to anyone. The centre buys seed from these groups based on demand only.

Box 11.2 Sweet potato vine growers.

'A few months ago our vegetables were photographed and now we have a contract to produce seed,' says Mr Seth Rakotondrazafy, surrounded by other farmers from Andranomanelatra village. The association to which he belongs has grown from strength to strength in recent years.

Madagascar's central highlands are famous for potato, sweet potato and vegetables. However, winters in Antsirabe, at 1600 metres altitude, regularly bring frosty nights, making the production of sweet potato vines nearly impossible.

But, in 2004, researchers from FIFAMANOR identified the nearby village where farmers were growing sweet potato on the warmer south flank of a hill. Farmers also made clever use of the waste water discharged by the TIKO milk processing factory to irrigate their fields.

Visiting their fields 6 years later, members of the Maitsoririnina association (meaning 'green, even in winter') proudly tell us they had already sold 20 tonnes of vines to the Ministry of Agriculture, via FIFAMANOR. They needed about 1 hectare, using three rounds of cuttings, following what they learnt from FIFAMANOR. They shifted to short cycle varieties that are high in demand: two orange-fleshed ones (*bora* and *mendrika*) and a white-fleshed one (*naveto*).

(*Continued*)

Box 11.2 (*Continued*).

In fact, the association's members are all family who cultivate their own land. The three women and five men mainly share crop production and market information with each other, and jointly decide on growing those varieties that are most desired that season. And they join forces when buyers from Tamatave or Tana (Antananarivo) visit them to buy in bulk.

Mr Seth, president of the association, tells us how all members take pride in growing crops. For their vegetable production they always buy seed from Semana that were conditioned and packaged in France. 'A few months ago we hauled our vegetables to the nearby market, where someone photographed them. He was impressed by the quality of our produce, enquired about our fields and took my mobile number. A few weeks later he called me, saying he worked for the seed multiplication centre (CMS) in Laniera, not too far from the capital, and asking to visit our fields,' recalls Seth. 'He asked the association to grow seed of petsaï, a Chinese cabbage.' The membership grew to 14 and a contract was established stipulating they will receive 3000 Ariary ($1.50) per *kapoaka* (a tin of condensed milk). Not sure if the contract will be profitable, they allocated a third of a hectare to this crop after having received source seed and training from the CMS Laniera.

The CMS staff has already informed them that next year they want to contract them to grow seed of cucumbers and courgettes, as part of a project with the World Vegetable Center (AVRDC). How a digital photograph can change people's lives!

Due to increasing problems with bacterial wilt, however, the centre is obliged to grow seed on fields that have no such disease history and as such has resumed production contracts with seed producer groups, albeit under very strict monitoring, assured by its network of extension agents.

Land. Of the 570 hectares spread over three stations, about 350 are cultivated, the rest being taken up by buildings or left fallow for potato seed production. No rent is paid to use the land, as the centre is still considered a public service.

Infrastructure and equipment. FIFAMANOR has well-kept offices, laboratories, screen houses, a seed conditioning unit and warehouses. While working with outgrowers, FIFAMANOR installed five potato and two cereal stores in rural areas. Since it stopped working with outgrowers, some of these have been used by cooperatives.

Links and partnerships. The centre collaborates with national (FOFIFA) and international centres, such as CIP and CIMMYT, from which it obtains lines (wheat) and clones (potato and sweet potato) to be tested and multiplied in Madagascar. The research department assists in the development of the national variety catalogue. FIFAMANOR is a member of FIOVA, the association of enterprises of the Vakinankaratra region, which organizes regular fairs.

Quality control. None of the seed produced by FIFAMANOR is certified by SOC, but this will change from 2011 onwards when the national seed catalogue is ready and more stringent official control procedures are in place. However, for potato, SOC doesn't have the facility to test for diseases, which is necessary for potato certification.

Potato and sweet potato plantlets grown *in vitro* in the laboratory are transferred to sterilized substrate in insect-proof screen houses to ensure healthy seed production. Samples are tested for viruses using the ELISA test (enzyme-linked immunosorbent assay) before potato seed is sold. Another big asset for the centre is its high altitude station in Tsiafajavona (2000 m), especially for potato breeder seed production. At this altitude very few aphids occur; aphids may transfer viruses.

For grain and forage crops, FIFAMANOR has its own quality control laboratory to test moisture content, germination rate and specific and varietal purity.

In vitro cultivation of potato and sweet potato is a first step to producing healthy plantlets.

11.6.3 Cash flow

Although salaries are paid by the government and infrastructure has been mainly obtained through donor support, the centre aims at becoming financially more independent.

Through its good links with research, foundation seed is either produced on-farm (for potato, sweet potato and wheat) or sourced from FOFIFA for the other crops.

Seed prices have remained relatively stable over the past 5 years, with fluctuations mainly caused by the unpredictable annual decision of the Ministry of Finance to consider seed as any other good and raise a 20% value added tax (VAT) on it.

Until 2006, FIFAMANOR contracted numerous farmer groups to grow potato seed, to whom they gave foundation seed on credit, but it almost entirely stopped doing that after contracts were not respected.

Potato foundation seed is sold at 1200 Ariary ($0.60), whereas seed potato is sold at 1000 Ariary ($0.50). One kilogram of sweet potato vines costs 440 Ariary ($0.22).

Table 11.9. Clients of FIFAMANOR.

	1995	2000	2006	2009	2015 (predicted)
Projects and NGOs	1	1	1	1	2
Independent seed producer groups	–	–	–	2	1
Farmer associations	3	3	3	2	1
Individual farmers	4	4	4	3	3
Contractual seed producer groups	2	2	2	–	–

Ranking assessment by senior management of seed enterprise, 1 being the most important. As different crops have different clients, the exercise was done for potato seed only.

11.6.4 Marketing

Clients for potato seed are mainly projects and NGOs, farmer associations and seed producer groups (Table 11.9). The system of contracting farmer associations to sell seed of potato and other crops helps to improve farmers' access: any farmer group can sign a contract with FIFAMANOR, pay 10% of the (favourable wholesale) price upon receipt and the rest a month later.

New varieties are promoted via FIFAMANOR's network of extension agents, who set up demonstration plots and nurseries in the villages where they work and live. Also, agricultural fairs and the national radio (RNM) are used, but for seed sales it generally does not work with the radio unless it faces problems selling.

11.7 Meva Seed Potato Growers' Cooperative

11.7.1 History

Androkavato, about 17km north-west of Antsirabe along an eroded and winding road beside terraced fields, is one of the villages that is blessed with a good climate at 1600 m altitude, fertile volcanic soils and a year-round water supply from a nearby source, which allow its farmers to grow three crops a year. Ten trucks a day come to buy ware potato, wheat, maize, carrots, cabbages, beans and a range of fruits, such as peach, pear, persimmon (*Diospyros kaki*) and *bibassier* (*Eriobotrya japonica* or Japanese plum).

Although some individuals were already producing potato seed in the 1980s, in 1991 seven farmers decided to form a seed producers' association. In 2003 they decided to register the Meva cooperative, as this allowed them to share any profit made between its members. They cleverly kept their association, as projects often give associations access to subsidized fertilizers. The membership had grown to 29, including seven women, all of whom contributed to paying the 100,000 Ariary ($50) registration fee of the cooperative. They then joined a union of 18 seed producer cooperatives in Vakinankaratra (VKMMV).

Already in the 1980s FIFAMANOR supplied them with foundation seed of popular potato varieties, the last one being *meva* (meaning nice, because of its pink flowers and prolific production), from which the cooperative borrowed its name. They used two of the potato stores established by FIFAMANOR with a capacity of 20 and 30 tonnes. In 1994, a third store was established with 30% of the costs borne by the association. The business expanded and its profit was used to build a fourth store with a capacity of 10 tonnes in 2003. Three years later, as their reputation in the region grew, they built a small shop for the sale of seed, fertilizers and pesticides (Table 11.10). The members elected two people to look after the shop. Half of the profit of the sales is for them, the other half for the cooperative.

Table 11.10. Seed produced (tonnes), Meva cooperative.

	2005	2006	2007	2008	2009
Potato	64	173	24[1]	23	8
Sales of agricultural inputs					
Fertilizer (tonnes)	1	27	10	4	8
Insecticides (l)	27	23	11	22	24
Fungicides (kg)	69	33	44	258	242

[1]Seed production contracts with FIFAMANOR ended.

Over the years, 18 potato varieties have been evaluated

in the demonstration plot set up by the FIFAMANOR extension agent. The cooperative retained five varieties of which it produces seed as well as ware potatoes. Farmers generally prefer varieties that keep well, which is another reason why *meva* is still the most popular one. They also grow seed of *marevaka*, which gives smooth and oblong tubers that are well appreciated by the urban market because it lends itself well to making chipped potatoes (French fries) and mashed potatoes. Ware potatoes of *marevaka* sell at a higher price than those of *meva*, but they are more sensitive to diseases. Other varieties grown include *diamondra*, *maharevo* (tubers of which have a similar diamond shape, but a pinker skin) and *marotia* (meaning loved by many). By 2007 they produced about 100 tonnes of potato seed on about 4 hectares and they have always easily sold all of their seed.

More recently, however, problems with bacterial wilt have meant that the cooperative could get only part of the foundation seed they used to buy. As they started to multiply their own seed, yields went down by more than 60%, so they realized the importance of renewing their foundation seed annually, if only they could. Customers became sad that the cooperative hardly sold any more seed, and so they started buying any seed they could find.

The cooperative also grows oat seed, but was unable to sell it all in 2009. Oat seed is mainly bought by dairy producers, but as these are mainly located near Antsirabe sales are not as easy as dairy farmers are reluctant to travel the 16 km of eroded road to the cooperative. 'Growing potato seed is demanding, but we are much surer that we will sell it all as we live in a potato growing zone,' says Zanalisoa Rasoanarivo. Apart from being one of the two people in charge of the shop, she is also treasurer of the association, union and cooperative.

Members of the cooperative have a passion for agriculture. To cope with recent changes in the potato seed sector, various members recently signed a contract with Malto to grow barley, which the company uses to produce malt for the Star brewery in Antsirabe (Box 11.3). Although members can grow any crop, they continue hoping that FIFAMANOR can solve the problem of access to potato foundation seed, as that is what the cooperative likes most and for which they have built a reputation.

11.7.2 Structure

Management. The cooperative has a president, a vice president, an adviser and a treasurer. Members started growing potato seed for FIFAMANOR, first as individuals and later on through contractual agreements whereby prices were fixed in advance and whereby foundation seed was provided on credit. Since 2006 they no longer work with such contracts. They buy their foundation seed from FIFAMANOR and sell to whoever is interested.

Land. The cooperative has no common land. All members grow seed on their own fields and profit is shared according to each member's contribution.

Infrastructure and equipment. The two potato stores and shop belong to the cooperative, whereas the two other stores belong to FIFAMANOR. All members have their individual equipment, including a knapsack sprayer.

Quality control. The treasurer's husband, Mr Dieudonné Randriamananjara, is an extension agent of FIFAMANOR who lives in the village. Although he has to oversee activities in three communes, he taught them a lot about quality seed production.

11.7.3 Cash flow

Members never took credit to produce seed potato. For the foundation seed the cooperative pays to FIFAMANOR 1200 Ariary ($0.60) per kg. While ware potato is sold at 600 Ariary ($0.30) per kg, the Meva cooperative sells its potato seed at 1200 Ariary ($0.60) per kg, irrespective of the client or the amount bought. This makes their quality seed just as expensive as FIFAMANOR's foundation seed.

11.7.4 Marketing

No advertising is done. The reputation of the Meva cooperative has spread through word of mouth only. Farmers are the key customers (Table 11.11). Potato seed is sold to individual farmers, coming from the village, neighbouring communes or even as far as Betafo, 37 km away. Quantities bought vary between 300 kg and 1 tonne (about 2 tonnes of seed are needed per hectare). NGOs and projects often place orders over the phone, but are less important and less predictable than direct sales to farmers.

Most clients return every season to buy seed, although the rate of return customers varies between seasons. Potato produced during the intermediary season (January/February to May/June) fetches a higher price, so 100% of the clients come and buy seed every year in

Box 11.3 Barley for the brewery.

Every year, the Malto factory turns about 4000 tonnes of barley into malt to supply the Star brewery in Antsirabe. For this, it works with 7200 farmers who grow barley in wintertime, in between two rice seasons. Malto used to establish contracts with roughly 500 farmer groups. The Meva cooperative was one of them. In order to avoid problems with group liability, Malto acted on farmers' own preference and now makes contracts with 6500 individuals and 225 groups, of on average three to four farmers. This seriously increased the repayment rate.

Malto provides its barley growers with quality seed, fertilizers, pesticides and technical advice through its 27 extension agents. Every year it needs 200 tonnes of seed. After having tested barley seed introduced from Tanzania and South Africa for their performance and brewing quality, four varieties were retained. These are multiplied by 250 trained seed growers, who receive 20 Ariary ($0.01) above the grain price. The seed is processed by Malto and distributed to its growers the year after.

As farms in the highlands near Antsirabe are small, the logistics are enormous. During harvest, Malto goes from farm to farm with a truck and a bus with 50 workers who weigh and bag the grain. The team is even fed by a caterer to increase efficiency. 'For two months and a half, we have a mobile factory,' says Daniel Couderc, director of Malto. Working with smallholders certainly has its challenges.

Table 11.11. Clients of Meva cooperative.

	1995	2000	2006	2009	2015 (predicted)
Individual farmers	1	1	1	1	1
Projects and NGOs	2	2	2	2	2
Farmer associations	4	4	3	3	3
FIFAMANOR	3	3	–	–	–

Ranking assessment by senior management of seed enterprise, 1 being the most important. As different crops have different clients, the exercise was only done for potato seed.

January. As production is much higher in the rainy season (October/November–February/March) and in the dry season (June/July–November/December), prices then are lower than in the intermediary season. Some of the customers who grow potato during these seasons recycle their seed, and so only about 40% are repeat customers at that time.

11.8 Challenges and Strengths of the Seed Enterprises

The 2008 global food crisis seemingly had little effect on Madagascar, which had anticipated the crisis and started to boost its domestic production to make the country self-sufficient in rice. But, while policies favoured stable rice prices, the cost of mineral fertilizers and fuel went up. Seed producers, reluctant to increase their seed price out of fear that farmers will stop buying altogether, saw their profits dwindle.

More disturbing has been the political crisis that started in February 2009 (and that had not been resolved at the time of writing this chapter in July 2010): companies closed down, the tourism industry collapsed, most bilateral aid was frozen, development organizations withdrew, and people's (already low) purchasing power dropped further. With projects and NGOs often being their principal clients, various enterprises have had problems selling some of their seed.

Madagascar faces frequent natural disasters on the one hand but has diverse agroecological zones where seed and crops for relief aid could be produced if the strong interest groups favouring rice imports could be managed.

On various occasions the authors were told that staff was given part of the locally produced rice seed to eat because it could not be sold, while other regions faced a shortage. Multi-stakeholder platforms could play a role in assessing seed demand and supply in a more systematic and transparent manner.

Seed relief agencies buy the cheapest seed on the market, in large amounts with little emphasis on quality. Until now, the legislation has not affected seed production and distribution. Even though the law forbids the sale of uncertified seed, companies only certify when customers ask for it. To raise seed quality, in 2010 the Ministry of Agriculture decided that the national laboratory would no longer test samples that are not from inspected fields. Most enterprises see this as a positive development which will reduce unfair competition.

Although most enterprises plan to extend their selling points, all act in isolation. Likewise, many projects have trained seed producer groups and strengthened their marketing skills, bypassing existing small traders and retailers with their long-established knowledge, skills and relations. Improving ties with other market actors, instead of trying to exclude them, would benefit seed producers and end-users just as in other value chain developments (e.g. Devaux *et al.*, 2010).

Although Fafchamps and Minten (1999) stated that most agricultural transactions in Madagascar take place without orders and without credit, many seed enterprises do take orders from projects, NGOs, farmer groups and at times even individuals. Few enterprises borrow money, but many sell seed on credit, although rarely to individuals.

The introduction of micro-finance institutions has raised many expectations (Dunford, 2009), but, with a 36% interest rate and the need for clients to present proof of identity (something poor people rarely have), Malagasy farmers often turn to local moneylenders whom they know. 'Many traditional social institutions exist, but we have always ignored them and tried to establish new economic ones, such as cooperatives, which often collapse once projects end. Seed distribution and marketing may benefit a lot by building on existing local institutions,' says Philibert Rakotoson, secretary general at the Ministry of Agriculture.

Apart from SCAA, the sweet potato growers' association and the Meva cooperative, all the enterprises described above were managed by (former) government people with long experience in seed. All enterprises showed adaptive management by diversifying their activities, often combining seed with crop production to speed up cash flow, and some enterprises started selling small seed packs to increase their outreach to farmers.

Despite the unstable political climate and unfavourable grain market and relief aid politics, many Malagasy seed entrepreneurs have managed to survive. Improving business support services and communication between all actors in the seed sector will be crucial to ensuring that the masses of poor people reap the benefits of quality seed.

References

Devaux, A., Andrade-Piedra, J., Horton, D., Ordinola, M., Thiele, G., Thomann, A. and Velasco, C. (2010) *Brokering Innovation for Sustainable Development: the Papa Andina Case.* ILAC Working Paper 12. Institutional Learning and Change Initiative, Rome.

Dorosh, P.A. (2008) Food price stabilisation and food security: international experience. *Bulletin of Indonesian Economic Studies* 44(1), 93–114.

Dunford, C. (2009) Credit unions and rural banks reaching down and out to the rural poor through group-based microfinance. *Enterprise Development and Microfinance* 20(2), 107–124.

Fafchamps, M. and Minten, B. (1999) Relationships and traders in Madagascar. *Journal of Development Studies* 35(6), 1–35.

IMF (2009) *Republic of Madagascar: Poverty Reduction Strategy Paper – Annual Progress Report.* International Monetary Fund, Washington.

MAEP (2007) *Recensement de l'Agriculture: Campagne 2004–2005, Tome III: Parcelles et Superficies des Cultures.* Ministère de l'Agriculture, de l'Elevage et de la Pêche, Antananarivo. http://harvestchoice.org/maps-etc/Madagascar_2005_Vol%203.pdf.

MAEP (2008) *Document de Stratégie Nationale Semencière.* Ministère de l'Agriculture, de l'Elevage et de la Pêche, Antananarivo.

MEI (2010) *Evaluation de la Filière Riz à Madagascar.* Ministère de l'Economie et de l'Industrie, Antananarivo.

Minten, B. and Barrett, C.B. (2008) Agricultural technology, productivity, and poverty in Madagascar. *World Development* 36(5), 797–822.

Minten, B., Dorosh, P., Dabat, M.-H., Jenn-Treyer, O., Magnay, J. and Razafintsalama, Z. (2006) *Rice Markets in Madagascar in Disarray: Policy Options for Increased Efficiency and Price Stabilization.* World Bank Africa Region Working Paper Series No. 101. World Bank, Washington.

12 Conclusions

Jeffery W. Bentley, Paul Van Mele and Robert G. Guéi

12.1 There is Life after Structural Adjustment

Despite massive privatization there still is a formal seed sector in Africa, including seed traders and growers and in some countries plant breeders and certification agencies. Few seed enterprises are still state owned, and even SONACOS in Morocco is now being privatized. Many of the seed enterprises are run by or employ agronomists or technicians who once worked for the public sector. But most formal seed enterprises are private, including medium-sized ones (like Premier Seed in Nigeria, Western Seed in Kenya and NASECO in Uganda). Many of these medium-sized companies produce seed through outgrowers, giving smallholder farmers a chance to earn cash income by growing traditional crops. All of the nine African countries we studied have small seed enterprises, sometimes run by a single farm family or an individual entrepreneur, like Mama Adama in Guinea or Mrs Coulibaly, who manages Faso Kaba in Mali.

12.2 Overview of the Book

In spite of their similarities, the nine countries reviewed in this book all have a unique seed history.

Cameroon (Chapter 3). The FAO set up farmer seed-producing groups in a remote, food insecure area, to provide seed locally. Some of the groups failed but some survived, especially the ones that were able to manage revolving funds. The groups were organized into a federation, which acquired source seed from a national research agency. These groups endured in part because they had few competitors and high demand from farmers in northern Cameroon for seed of new, short duration rice, maize, sorghum and millet varieties that help them cope with a changing climate.

Nigeria (Chapter 4). Private companies are organized in clusters around cities like Zaria, which have resources and people that companies need, such as universities and foundation seed suppliers. The companies see themselves locked in unfair competition with state agencies called ADPs, although the ADPs tend to produce the crops that are least profitable for the companies, such as rice and sorghum, while companies rely more heavily on hybrid maize and vegetables, although all of the companies do produce various kinds of crop seed. The federal government buys enough seed to seriously distort the market, which is one reason why multinational seed companies have come to Nigeria only to leave again. The Nigerian government's National Agricultural Seed Council does a competent job producing and distributing foundation seed and conducting quality control.

Mali (Chapter 5). Faso Kaba is a company started by the wife of a plant breeder, who used experience earned in the USA and her husband's help to start a business, buying and selling seed. AGRA has asked Faso Kaba to produce seed as well, which it did after it acquired land and recruited an agronomist and an experienced seed technician. Niégué Farm is a former state-owned farm, privatized and turned over to a group of agronomists, who used its land, tools and irrigation system to produce abundant rice seed. COPROSEM and Nipagnon cooperatives both produce sorghum and maize seed, whereas Nipagnon also grows seed of rice, cowpea and other crops. Nipagnon uses radio advertisements to promote its seed, and COPROSEM announces its local seed fairs over the radio. Both cooperatives supply local seed traders, but Nipagnon also sells to customers from neighbouring countries at the national seed fairs organized by the Ministry of Agriculture. Some villages in the arid Dogon country have acquired enough fame for the quality seed of their early maturing pearl millet landraces to attract the attention of international NGOs and local traders, including Faso Kaba.

Guinea (Chapter 6). Mama Adama and Sherif are really local. Sherif had capital support and training, but Mama Adama produces and sells seed entirely on her own. The Comptoir Agricole is a flexible seed company now riding out the world financial crisis by putting its employees on straight commission instead of salary. The Union of Cooperatives produces potato and cereal seed on the Fouta Plateau. Hundreds of highly motivated women work together, supervised by government extension agents, who manage the Union. El-Hadj Sow is a trader, inspired by his visit to France, who started growing potato, rice, maize and cowpea seed largely on his own, and he made it work. Certification services are nascent in Guinea and often dispensed with, creating an opportunity for small, private enterprises to produce quality seed, but also blocking entrepreneurs like El-Hadj from exporting seed to neighbouring countries.

The Gambia (Chapter 7). Since the 1990s, Gambia Horticultural Enterprises has continuously invested in marketing and has grown from a small office in the owner's home near Banjul to the country's largest farm supply company, selling seed, agrochemicals and machinery, and even exporting some products to Europe. The lack of a seed certification agency created a space for an ambitious farmers' association, the Jambur Kafo, mostly women. They have strong ties to government researchers, who give them seed and advice. These farmers became interested in seed while doing participatory varietal selection (PVS), but they were already a self-organized group before that. They make deft use of radio and TV to advertise nationally. The other case is Mr Dembo, an enterprising émigré, who returned to The Gambia, got seed and advice from the Jambur Kafo and used his own capital to start a thriving rice seed farm. The Jambur Kafo and Mr Dembo are both near the capital city, and have close ties with research, extension and media.

Morocco (Chapter 8). The one big state seed company in Morocco, SONACOS, defied structural adjustment and produced seed of wheat and other crops that private enterprises might have rejected. While this state company has received public support to produce seeds, we cannot say if it is economically efficient or not. But a public agency should be judged a success if it renders the services it is assigned and remains free of corruption and mismanagement. By that standard, SONACOS has been successful. For many years it put quality seed into the hands of smallholders. However, overall agricultural output in Morocco has declined. While this cannot be blamed on

SONACOS, it is possible that declining demand for seed (and a feeling that change was finally in order) has motivated the Moroccan government to start to privatize agriculture and seed.

Kenya (Chapter 9). The diverse seed sector includes a medium-sized private seed company that produces a lot of maize seed (OPV and hybrid), like some of the Nigerian companies. There is a new farmers' cooperative of soybean producers, which may or may not survive without the care of an institution. Sungus is a family farm that used a bit of land, some creativity and hard work to benefit from a German seed course to become a professional producer of seed potatoes.

Uganda (Chapter 10). Building on an NGO initiative, a Belgian agronomist started a company in an abandoned cotton ginnery in a village with bad roads and no electricity, eventually producing two or three times as much seed as other companies around Africa. In another case, a group of women who started as outgrowers are now producing rice and maize seed and dream of starting a company. A third enterprise started as a farmer field school (FFS) in 1999 which selected disease resistant bean varieties and ended up training and supervising various bean producers' groups, selling small packs of seed through a network of agro-dealers. As measured by longevity this enterprise is successful, although they still have support from CIAT, an international research centre.

Madagascar (Chapter 11). Valy Prod Sem is a bean seed company that grew out of an experience with a privatized government seed centre, and is linked to communication and entertainment businesses belonging to the same owner. Sakay is a privatized government seed centre that sells mainly to projects, but mobile phones now allow it to increasingly sell directly to farmers. SCAA is the largest government rice seed centre, which has recently come under the management of a private entrepreneur. Andri-Ko is a private company, started by a family with experience in the Ministry of Agriculture. FIFAMANOR is a public enterprise that evolved from a project funded by the government of Norway, and is now striving to recover more of its costs, mainly through foundation and quality potato seed production and marketing. MEVA is a cooperative where the high altitude allows some growers to produce high quality seed potato for other farmers.

These diverse, African seed enterprises serve a range of crops, giving each enterprise its own needs and challenges. The next section describes some of the unique challenges of enterprises dealing with four crop types. The third section describes business models, which are shaped by the kind of seeds they produce.

12.3 Unique Challenges of Four Crop Types

The four crop types are open-pollinated, self-pollinated, legumes and vegetatively reproduced. Open-pollinated crops, such as maize and pearl millet, cross easily, exchanging pollen with their neighbours, making it more challenging to maintain pure varieties. Seed of self-pollinated crops is usually easier to produce, and may be less profitable, especially for field crops. Legumes (most of which are also self-pollinated) are perhaps even less profitable because of the varieties' very slow deterioration rate (Section 2.2). Vegetatively reproduced crops have bulky planting material instead of light, portable seed, and their stems, corms or tubers are perishable. Each of the four crop types has its own seed management implications, of which a thorough overview is given in the book *Farmers' Seed Production* (Almekinders and Louwaars, 1999).

Open-pollinated crops (maize, sorghum and millet). Seed of open-pollinated crops must be kept isolated from other varieties of the same species, or the varieties will cross with each other. Some varieties need to be a kilometre away from other stands of the same crop. So a smallholder seed producer can only grow one variety and it must be far from other farms. Or the seed producer can plant earlier or later. Farmers may need training to learn to produce pure varieties of open-pollinated seed.

Self-pollinated (rice and wheat). Seeds of self-pollinated cereals are the easiest to save on farm without losing their genetic identity, making them appear less attractive to private companies, so governments often believe they have to produce seed of these crops, even though this may actually not be the case. This book has shown various examples of private seed enterprises that do grow rice seed, including a few that only produce rice seed. Under certain conditions, it may be profitable to grow rice seed, e.g. in high production areas (Mama Adama in Guinea, SCAA in Madagascar, Niégué farm in Office du Niger) or in countries like Uganda where grain markets are protected or rapidly expanding, and when new opportunities arise (relief aid, export or introduction of a new variety.

Legumes (beans, soybeans, cowpeas and groundnuts). Legume grains are important to the communities, but large seed companies have scant interest in grain legumes, because economic returns are low or because legumes are more difficult to harvest with machinery than cereals. Seeding rates are higher for legumes than for maize. One may plant a hectare of maize with 30 kg of seed, compared with 100 kg for beans. This increases farmers' incentives to save their own legume seed, which is easy to do since legumes are predominantly self-pollinating.

Vegetatively propagated (potatoes). Although ware potato (for food) can be grown at low elevations in tropical countries, seed potato cannot. Seed potatoes can carry disease and even insect pests, like tuber moths. In the tropical lowlands conditions are usually more favourable for pathogens and insects, giving farmers a compelling reason to buy seed, creating a constant demand for seed potato growers who can provide healthy seed. The seeding rate is so high (as much as 2 tonnes per hectare) that potato growers are sensitive to price and usually prefer uncertified seed, to keep costs low. Other crops like sweet potato, cassava and banana are rarely taken up by seed enterprises, and hence hardly feature in our book.

12.4 Business Models

Crop type influences supply and demand for seed, but leaves plenty of room for individual variation. This section presents business models that reflect the nature of the people who run the seed enterprises, including farmers, farmers' associations, small seed companies (including traders), medium-sized companies and public agencies. As Harper and Tanburn (2005) observe for agriculture in general, seed enterprises in Africa rely on other small enterprises for services (skills, information, inputs, storage, marketing, equipment, transport, repairs and maintenance). Links with foreign enterprises are also beneficial when they exist (e.g. Gambia Horticultural Enterprises; Vali Prod Sem in Madagascar). Government-organized seed fairs, like those in Mali, have helped to create national and international demand and opportunities for various types of enterprises.

12.4.1 Farmers

Bello Yarima in Nigeria is an individual farmer selling open-pollinated maize and sorghum seed (Box 4.2). Mr Yarima also sells legumes, vegetables and other seed he grows himself. There are actually few examples of this type in our study, but there are probably thousands of informal seed producers like him across Africa. The authors only met Mr Yarima by accident. In the arid northern part of Mali, a few Dogon villages have acquired a good reputation with their early maturing millet landraces. Landraces are evolving, and are also genetically more diverse than improved varieties (Brush, 2004) but the Dogon maintain quality and uniqueness through traditional community rules and norms.

El-Hadj Sow in Guinea maintains high seed diversity (potato, rice, maize and cowpea). He is well trained and has strong links with research, extension and the private sector. He uses rural radio to promote his Nerica seed.

But Mama Adama, also in Guinea, may be a more typical case of a farmer seed producer. She specializes in seed of local rice varieties. She has no ties with research and extension, and advertises by word of mouth, using trusted, repeat customers. Mr Dembo in The Gambia produces rice seed (for a new variety which is in high demand); he grows other crops too, but not for seed. That is a difference between farmers and seed companies. Farmers diversify into other crops and livestock. Seed companies diversify into other types of seed and other agricultural inputs.

Some farmer seed producers establish themselves and prosper through their strong political ties or because they are solid members of the community. Mama Adama is the wife of a respected imam. El-Hadj Sow is the resident of his prefectural Chamber of Agriculture. El-Hadj has benefited from government support, but the other farmer seed producers have not. They invest in rice seed with their own money, suggesting that credit or financial support is not always essential to maintain a small business, but an enterprise often needs credit to expand. If financial products were better tuned towards agriculture and the seed sector (e.g. lower interest rates, repayment schedules that matched the harvest season), there might be more growth in African agricultural economies.

Most smallholder seed producers do not certify their seed. Certification adds to the costs and technical requirements, and demands contacts with the formal sector. Seed certification is a barrier for small-scale seed producers.

El-Hadj Sow in Guinea and Sungus Seed Potato Enterprise in Kenya sell their seed potato locally, straight from their farm. Sungus is a diverse family firm, with other crops, a dairy, even a hotel, but they do not produce seed for any crop other than potato. Neither El-Hadj nor Sungus have had bank loans; they work with their own funds and are not looking for loans. For potato, more than for other crops, a source of clean seed is crucial. Sungus in Kenya gets source seed from the national research agency (KARI).

12.4.2 Farmers' associations

A few associations produce OPV maize seed, such as COPROSEM in Mali and the farmer groups in Cameroon. It is easier for groups to grow self-pollinated crops such as rice or beans. For example, The Gambia's Jambur Kafo and the Sayem Seed

Production Union in Cameroon produce and sell rice seed. They have strong links with research and extension, which gives them the technical and marketing training they need. They use creative advertising. All of these associations diversify, usually by producing grains, vegetables or livestock.

Associations can easily sell directly to farmers, but some of them also sell seed to projects and NGOs, which are important seed buyers and may keep many seed enterprises in business. Some farmers' associations own some machinery, but, true to their smallholder heritage, they are able to keep overheads low.

Most seed producing associations had support from donors to get started or they received much support from government. They needed start-up capital. Many also had business development support (smart subsidies) and institutional strengthening of the group leaders. They usually had good access to foundation seeds.

Most associations or cooperatives produce high volumes of seed but do not certify it, relying instead on consumer trust of their name. Some use small seed packs, targeting the rural poor. Associations need easy seed regulations or at least ways to avoid confronting the seed laws. They usually cannot go through the whole process of certification, which makes their seed too expensive for the local farmers who are their customers. All seed-producing cooperatives and associations in Mali have received project (PAFISEM) support to train their members and to have their seed certified. The future will tell how many of them will survive now that the project has come to an end.

Many associations produce grain legume seed, possibly because they face little competition from private companies. For example, the Nyamabale Bean Seed Producers' Association in Uganda and the Mumias District Federation of Soybean Producers (MUDIFESOF) in Kenya are community enterprises that keep costs low and prices competitive, because they know they are competing with the grain market and farm-saved seed.

The Cereal and Potato Seed Producers' Union in Guinea and the Meva cooperative in Madagascar are the only two cases in our book of an association producing potato seed. The other cases are family farms (Sungus in Kenya and El-Hadj in Guinea). The Union is managed by qualified agronomists and the Meva cooperative has regularly received technical support from research and extension. All four of these enterprises sell much of their seed locally to reduce transportation costs of the bulky seed. None of them certify their seed, which also helps lower costs. Their reputation is their seal of quality.

12.4.3 Small seed companies

Traders such as Faso Kaba (Mali) and Sherif (Guinea) can easily market OPVs, if they do not have to grow them. They can also buy and sell seed of other types. Some traders, like John Kyoma in Uganda, grow small fields of the seed they promote, just to learn about the variety and to be able to give better advice to their clients.

Small seed companies may have their own land but they use outgrowers for all or most of their seed supply. Small seed companies work hard to keep the trust of outgrowers and farmers. They try to make farmers their main clients, and try to sell less to the government, because selling to the government can lead to crushing debt. The healthier companies also avoid bank loans with high interest rates. They need to make a profit to survive, but they grow slowly and steadily, often on their own funds, keeping their overheads as low as possible, for example by renting their offices or their processing equipment instead of buying it, or by doing without cars, air conditioners or other expenses.

For many small companies, their core seed product is hybrid maize seed or imported vegetable seed. Few rely mainly on rice or wheat seed, where the company faces stiff competition from farmer-saved seed. Companies that do sell grain seeds may receive government support or use a vertical value chain approach (such as Terratiga in Nigeria), selling to grain producers themselves linked to the market through rice milling. Some such as Andri-Ko (Madagascar) have a restricted market area, but they forge solid links with research and development. Small does not mean sloppy. These companies have well-educated, experienced managers who often started in larger companies or the public sector, and they know their business.

Small seed enterprises often start by selling imported vegetable seed and later diversify into grain crop seeds.

12.4.4 Medium-sized seed companies

Medium-sized companies mainly produce seed of modern high yielding varieties, the result of formal research. Most of this seed is certified and is sold through wholesalers and retailers, rather than directly to farmers, although trends are changing. These companies have seed in their variety portfolio that is complex to produce, like hybrid maize, or they import vegetable seed, which farmers generally need to buy every season. The companies do not specialize in bean or rice seed, because it is difficult to compete with farm-saved seed. However, these companies often do sell some rice, sorghum, OPV maize or legume seed, as a way of diversifying and keeping a certain visibility in the market and being able to take advantage of big sales to government and NGOs.

Premier Seed in Nigeria has its own research department. This is unusual, even for a medium-sized company, but all the companies have experienced, capable managers. Most or all of them own their seed factories, some of which are endowments from earlier projects or the public sector, while others have invested their own money in equipment. Most of the machinery inherited from parastatals is now 20 or 30 years old, is often too big, and even with devoted maintenance cannot last for much longer. Medium-sized companies distribute nationally, through seed distributors or agro-dealers or through their own outlets, and to some extent on the international market, such as NASECO in Uganda.

Medium-sized African seed companies use outgrowers, who are often smallholder farmers. This passes the risk of crop failure on to the outgrowers and is a tacit admission that smallholders are more efficient at producing cereal and legume crops than are private companies (see Wiggins, 2009, on the hardiness of African smallholders). Few companies buy land to grow their own seed.

As with the small companies, medium ones avoid bank loans whenever possible. Only entrepreneurs that have a long track record of proven success have been able to negotiate their terms and conditions for bank loans. The high transaction costs and interest rates of bank loans in Africa make formal loans an unbearable burden for most enterprises, although there have been efforts in Mali and Uganda to address this. Entrepreneurs with a long track record may negotiate their terms and conditions for loans, like GHE in The Gambia.

Medium-sized companies understand their national context perfectly and know how to build trust with outgrowers and farmers. They have close links with extension, research and development. They often employ former staff of international research centres, and maintain collegial relations with national and international scientists.

Value chain model. Some medium-sized enterprises produce one variety dedicated to one big customer, such as Terratiga in Nigeria producing sorghum seed for farmers selling to a brewery.

12.4.5 Public agencies

Public agencies, like Nigeria's ADPs or SONACOS in Morocco, are well-structured, state-owned agencies. They often produce large amounts of rice, wheat, grain legumes or other self-pollinated seed which is not overly attractive to private companies. However, SONACOS also produces seed of sugarbeet, potatoes, vegetables and other crops which could be grown by the private sector. SONACOS specializes in seed and until recently held a monopoly on it, but Nigeria's ADPs are state extension agencies that sell seed as just one of their activities, but one which is popular with rural voters.

The Moroccan government made all the initial investments in SONACOS, including the physical plant and staff training, before gradually withdrawing in favour of private management. Yet privatization is only partial; SONACOS still depends on low interest loans and government subsidies to keep seed prices affordable to farmer customers.

Like private companies, these public agencies rely on outgrowers to produce the seed. The agency trains and organizes the outgrowers, collects, processes, stores and distributes the seed. With nationwide coverage, these agencies can help implement government policy to encourage improved varieties or certified seed for staple crops.

Public enterprises may well discourage private companies. Until recently SONACOS held a monopoly on seed. This is now being liberalized and it is too soon to tell what will happen to newly emerging companies in Morocco. In Nigeria, the ADPs have always competed with private companies, unfairly so in the eyes of the companies, but in truth the ADPs have also helped the companies by promoting hybrid maize and new varieties.

The state. Even governments that don't produce seed are involved in it through credit, registration, quality control, and seed certification and by providing new varieties and foundation seed. In Uganda and Mali, effective joint donor and multi-stakeholder financial coordination systems have favoured seed companies. Government support also includes tariffs on grain imports, subsidies, infrastructure, research (e.g. plant breeding), extension (promoting varieties), price support, and facilitating markets for seed, farm produce, fertilizer and other inputs. Some governments, like those of The Gambia and Guinea, are less involved but at least provide foundation seed and help promote modern crop varieties.

Certification and organization. Certification is a government responsibility. However, certification is not always worth the time and money it takes to do it, especially when mechanisms to trace fraudulent sales and legal systems to pursue them are lacking. Many of the successful enterprises in this book, especially the smaller ones, do not certify their seed. It is often more important to screen for plant diseases than to worry about genetic contamination. And, if the state ensures good foundation seed, certification may be less important.

Farmers can produce and sell common seed on their own, but, if they join the formal seed sector or want to tap into a network of agro-dealers, they profit from being organized in groups and receiving support from the public sector, NGOs or even private companies.

12.5 What Makes Seed Enterprises Tick

Equipment and infrastructure. Successful enterprises have access to good equipment, like seed processors, even if it is rented or borrowed. Processing equipment, storage facilities and roads (e.g. Nigeria, Mali and Cameroon) may have been left by government or projects before structural adjustment. Some companies like the small Chinese equipment. The old US or European project equipment 30 years ago was too big.

Larger enterprises have access to machinery that farmer associations lack.

Policy support. The National Bank in Mali explicitly supports the seed sector and this can be replicated in other countries.

Self capital. In Cameroon, the FAO insisted that farmer seed-producing groups use a revolving fund. The project gave the groups seed, fertilizers, pesticides and bags for seed packing, and they repaid the cost back to themselves to start their revolving fund so they could give themselves loans. The groups that survived were the ones that managed to keep their revolving fund.

Frugality. Money is important, but managing cash flow is more important than making a big investment. Successful companies of all sizes have lean overheads and avoid loans as much as possible.

Isolated enterprises are at a disadvantage. Professional organizations are important to defend their rights and improve prices, marketing, seed rules, controls, etc. Seed trade associations are often the result of seed sector support by donors or government. Even so, they are a forum where people discuss issues and decide what to do as a group. Seed enterprises need links with government, international centres or other companies for source seed, new varieties and technologies.

Contacts. No matter what you know about seed, who you know may matter even more. People who have been in politics or the ministry of agriculture have a head start if they want to start a seed enterprise. Many of the ones in this book started with some contact through a ministry of agriculture. Others were initially involved in seed related activities with an NGO or international research organization.

Diversity. Successful enterprises diversify. Most companies sell hybrid maize and vegetable seed, besides OPV seed. Successful agro-dealers sell fertilizer and other inputs besides seed. Successful smallholder seed producers still tend other crops and livestock. Many enterprises produce seed in the rainy season and something else in the dry season, often vegetables (not vegetable seed, just the vegetables).

Knowledge. The successful companies were started by people who knew how to manage a business. Good managers start small and build up creatively, without over-extending themselves. They train their staff, invest in them and try to maintain them, like the Nigerian ADPs and Premier Seed, which have stable staff. These enterprises train their outgrowers and distributors, and they treat their outgrowers with respect and honesty.

Not all seed enterprises know much about seed when they get started. Some, like Mr Dembo in The Gambia, were even new to agriculture. But they are all fast learners, and successful seed enterprises are sponges for new information.

Research. Successful enterprises invest in research and development, including market research. Few of the medium-sized companies breed new crop varieties, but many have research plots where they test new varieties. The farmers who produce seed are born innovators who enjoy trying new things. All seed enterprises are innovative by nature.

Decentralization. Certification services in Cameroon and Madagascar are being decentralized. Having only one central laboratory makes for delays, especially if regions are poorly connected. Several regional labs can process samples faster. The national laboratory oversees the regional ones if a government or the seed industry supports certification.

Quality. Successful enterprises build trust and reputation for quality products, like healthy seed. They build up strong internal support for inner quality control systems.

Certified seed is overrated. Many successful enterprises do not certify their seed, and none of the seed potato enterprises do. Insisting that all seed be certified is foolhardy if the government cannot certify everyone. Certification always implies a cost and, if not subsidized, makes seeds more expensive. Certification is good if it improves quality control. But the transaction costs and the social contacts needed for certification favour larger enterprises and lock out smallholders. Truthfully labelled seed may be a more feasible way forward, with more emphasis on branding and trust in the source.

Marketing strategies include radio campaigns and TV ads, demonstrations, seed fairs, billboards and field days, proper pricing, small packaging and labelling. Successful enterprises

Seed enterprises increasingly invite the media to broadcast farmer field days and promote their products and services.

keep proper records of their activities so they can see trends and make plans. And they sell to farmers, not just to government and NGOs. Unfortunately, government distortions may inhibit marketing. The Kenyan government has a draconian prohibition on advertising informal seed, even if it is of top quality.

Some successful enterprises (especially the medium-sized ones) provide backup support for farmers who buy their seed. They invest in small packaging (such as half kilo to 1 or 2 kg seed bags), which reduces fraud and makes seed accessible and affordable to farmers.

Successful enterprises cooperate. They join together in associations, try to protect the market from counterfeiters, share or rent equipment to each other. They may use the same research centres and source seed and some cluster around the same cities. The next section offers suggestions for helping seed enterprises, not just as individuals, but also as systems made up of enterprises and their support networks.

12.6 How to Encourage Seed Enterprises

12.6.1 For donors

Strengthen the financial sector to make debt bearable, by making appropriate loans for seed enterprises, which have to spend money for almost 2 years before they start making money. Provide loan guarantee funds for agriculture. Make loans available to small seed enterprises as well.

Support companies with equipment for seed processing, storage and training. Support professional bodies, like seed trader associations and agro-input dealer associations, at least during the start-up phase. Facilitate linkages between actors in the seed and commodity value chains, e.g. through multi-stakeholder platforms.

Support new technologies, irrigation, roads and electricity. Help the countries to do effective and inexpensive quality control and certification. General agricultural development will also benefit seed producers by increasing the demand for seed.

Support variety development and selection with farmers (e.g. the sorghum varieties by COPROSEM

Involving farmers, seed dealers and variety release agents in PVS speeds up the development, release and dissemination of better varieties.

followed PVS with ICRISAT; the beans in Uganda with CIAT; and rice seed in The Gambia followed PVS with AfricaRice). Strengthen each country's ability to produce breeder and foundation seed. This is a public sector task, especially when meeting smallholders' demands for seed and crop varieties.

12.6.2 For governments

Governments and donors can give loan guarantee funds to several commercial banks. In Uganda donors put money in more than one bank so they compete to give farmers the best services.

Cut taxes for seed companies and cut duties for importing machinery, inputs and seed. Provide training for seed enterprise staff. Train farmers in new crop technology and develop high quality training materials that can be used by many small-scale service providers. Have degree training for seed technologists. Develop curricula at universities. Subsidize airtime for local and national media for agricultural programmes.

Protect grain markets from cheap, subsidized imports.

Avoid buying large amounts of seed purchases to give free to farmers, which distorts the market. Governments should pay on time if they buy from seed enterprises.

Private companies can manage hybrid maize and vegetables. Farmers can manage bean and cereal seed. But they need help getting new varieties. By allowing truthfully labelled (uncertified) seed to enter the system, small enterprises can more easily market their seed beyond their immediate vicinity (by making their seed legal and allowing them to use existing agro-dealer networks). Small enterprises need equitable access to foundation seed and technical support.

Do not use regulations as barriers to entry for new and farmer-owned seed enterprises. Have different standards and regulations for small and medium-sized enterprises. Create appropriate small enterprise registration. Seed certification should be for quality control, not to create monopolies. Governments must prosecute fraud and hold cheaters accountable for bad quality.

12.6.3 For research

Research agenda. Develop appropriate, new high performing varieties. Do research on seed technology and comparisons of seed from different sources, including landraces, under different regimes. Study farmers' seed systems and enterprises. Study farmer demand more carefully, and involve farmers in setting the research agenda.

Involve farmers and end-users like consumers, agro-dealers and the seed companies in varietal testing. Also involve the people who allow a new variety to be released in the varietal testing so that they already know the qualities of the variety to speed up the release.

Certified seed is grown from foundation seed, which is grown from breeder seed, which is provided by the plant breeders. There is some suggestion that, after releasing a new variety, breeders are more interested in creating their next hit variety than in the thankless work of endlessly rearing out their older ones. Yet research centres must provide the highest quality breeder seed all the time to keep the varieties from getting mixed, and lost.

12.6.4 For NGOs and development agencies

Buy standard seed some of the time, if that is legal, and do not insist on all certified seed. Seed health is more important than certification per se.

Recover at least the cost of the seed. You can subsidize seed but don't just give it away for free. Sell the seed no lower than the grain price.

Stop buying so much seed. It distorts the seed market and creates instability.

12.6.5 For financial institutions

As mentioned above, make appropriate financial products and services for agriculture (Harper, 2005), and seed enterprises in particular. Loan officers need to be trained on the workings and needs of seed enterprises to be able to judge their risks and the services and products they need.

When creating loan guarantee funds, also provide technical assistance. Enterprises that require the guarantee from a fund are usually too risky for banks and so are 'non-bankable'. Training (e.g. on financial literacy or financial management) can make enterprises more likely to repay their loans.

Seed enterprises can use their commercial relationships with agro-dealers or farmers to increase access to finance. Assured sales of seeds demonstrate lower risks for financial institutions.

Financial institutions can facilitate the development of specific value chain finance schemes, such as warehousing.

Banks can enquire about the seed enterprises' strategies, linkages and clients before working out specific client-tailored products and services to ensure higher payback rates.

12.7 Conclusions

Old stereotypes die hard. As late as 2006, experts on African seed were writing that 'Seed markets in the region are small and highly fragmented; with closed national markets dominated by a few international companies and parastatals' and still reeling from the market and institutional failures caused by structural adjustment (Minde and Waithaka, 2006, 2–3).

Since the 1980s, African seed enterprises have come to life and grown. We have seen that seed certification may keep smallholders out of the seed business, and that there are some farm families and farmer groups successfully producing seed, although most of them still have some technical support from outsiders. We have met agro-dealers who are not usurious demons, but honest people who make seed flow, like the Comptoir Agricole, and respected individuals, like Mama Adama, who sells seed to her neighbours. Medium-sized companies often work hard to respond to increasing demands for seed and benefit from a well-organized network of agro-dealers.

For many years seed scholars have suggested that enterprises need a regular infusion of new varieties to stay viable (see Chapter 2). But it is not always true. Mama Adama has been selling the same variety for years to a limited number of clients. Niégué Farm in the Office du Niger has about five popular rice varieties that keep selling. Bringing a new variety on the market offers a marketing opportunity, but African economies have fewer competitors (and higher transport costs) than elsewhere, leaving local markets more open to a few actors. Perhaps seed scholars were influenced too much by Western perceptions of competition.

Another old problem is that many crop varieties are poorly adapted to farmers' conditions. Participatory plant breeding was created to address that – and seems to be working. Plant breeders are much more likely to involve farmers in their research than other agricultural scientists, and plant breeders are now taking many characteristics into account (e.g. taste, colour, disease resistance), no longer focusing only on yield.

New African enterprises. The fresh new type of enterprises described in this book have closer contacts with farmers than the old parastatals, and on a wider scale than scientists will ever be able to obtain. This is another reason why the public sector should strengthen the private sector. Future breeding should build more on local knowledge, not just of farmers organized in groups, but of seed companies, dealers and retailers, which will speed up the number of acceptable varieties.

The public sector is crucial to create a formal seed sector. That is not merely true by definition. In each of the nine countries the public sector (including international research centres) contributes at least some functions, especially basic plant breeding, which the private sector will never do.

Emerging entrepreneurial (human) characteristics. It is often assumed that entrepreneurs make it because they take risks, but, in the cases of African seed, it is more a matter of having a good sense for opportunities and for avoiding needless risk. Just like the examples described in the book *Africa's Greatest Entrepreneurs* (Makura, 2008), the seed entrepreneurs described in our book are loyal to their clients and hard-working and exude a real passion for the work they do. But they are generally more careful when it comes to risk-taking. The seed enterprises spotted opportunities and followed them with a stepwise, thoughtful approach, avoiding risky loans, trying it out at a small scale, much like farmer experimentation but with bigger stakes. As enterprises grew, they also became more risk averse as a failure would have more severe consequences.

The successful enterprises are also grounded in the community, whether that is a village or a region. Almost all of the entrepreneurs are from the area where they work, but they are all hands-on people, the kind who are in the seed factory or with the outgrowers every day, making sure that things are done right.

We have been fairly generous with the word 'successful'. We did not audit the enterprises financially or ask for privileged information. We used survival as our main yardstick. Anyone who can make a business, an agency or an association last for several years running is doing something right.

Structural adjustment's flaws are well-documented, including debt, unemployment, deterioration of social services and a general failure to deliver on its promises (e.g. Sachs, 2005; Meredith, 2006). But the parastatal seed corporations that the IMF broke apart in Africa may have been dysfunctional. The Moroccan case is like an experiment, where one African country held on to its parastatal seed company, but Moroccan agriculture has failed to grow and the government is now privatizing the public seed company of its own free will.

The cases in this book show that, whatever the merits and flaws of the parastatals, they were staffed by some educated, hard-working people, who went on to create successful small and medium companies after structural adjustment. The African seed enterprises described in this book emerged after 1980, most after 1990, from the wreckage of structural adjustment.

Structural adjustment may not have harmed smallholders. It reversed unfavourable terms of trade for agriculture; actually the 1980s were a period of accelerated growth for agriculture in Africa. African agriculture only looks stagnant when one includes the disastrous decade of the 1970s, before structural adjustment (Wiggins, 2009).

What is clear is that the retreat or advance of the state creates or closes spaces for seed enterprises. This book has explored how these spaces change and how seed entrepreneurs adapt to them.

References

Almekinders, C.J.M. and Louwaars, N.P. (1999) *Farmers' Seed Production: New Approaches and Practices.* Intermediate Technology Publications, London. 291 pp.

Brush, S.B. (2004) *Farmers' Bounty: Locating Crop Diversity in the Contemporary World.* Yale University Press, New Haven. 327 pp.

Harper, M. (2005) Farm credit and microfinance: is there a critical mismatch? *Small Enterprise Development* 16(3), 58–67.

Harper, M. and Tanburn, J. (2005) *Mapping the Shift in Business Development Services. Making Markets Work for the Poor.* ITDG Publishing, London. 181 pp.

Makura, M. (2008) *Africa's Greatest Entrepreneurs.* Penguin Books, Johannesburg. 291 pp.

Meredith, M. (2006) *The State of Africa: a History of Fifty Years of Independence.* Free Press, London. 752 pp.

Minde, I. and Waithaka, M. (2006) Rationalization and harmonization of seed policies and regulations in Eastern and Central Africa: effecting policy change through private–public partnerships. Paper read at the International Association of Agricultural Economists Conference, Gold Coast, Australia, 12–18 August 2006.

Sachs, J.D. (2005) *The End of Poverty: Economic Possibilities for Our Time.* Penguin Books, New York. 397 pp.

Wiggins, S. (2009) Can the smallholder model deliver poverty reduction and food security for a rapidly growing population in Africa? Paper for 'How to Feed the World in 2050', FAO, 24–26 June.

Acronyms and Abbreviations

ACOD	Association Conseil pour le Développement (Mali)
ADC	Agricultural Development Corporation (Kenya)
ADP	agricultural development project (Nigeria)
AELP	Africa Emergency Locust Project
AFD	Agence Française de Développement
AfDB	African Development Bank
AfricaRice	Africa Rice Center
AFSTA	African Seed Trade Association
AGRA	Alliance for a Green Revolution in Africa
AMMS	Association Marocaine des Multiplicateurs des Semences (Morocco)
AMPROSEM	Association Malagasy pour la Promotion des Semences (Madagascar)
AMSP	Association Marocaine des Semences et Plants (Morocco)
ANPROCA	Agence Nationale de Promotion et de Conseil Agricole (Guinea)
AOPP	Association des Organisations Professionnelles Paysannes (Mali)
APEK	Association pour la Promotion Economique de Kindia (Guinea)
APEP	Agricultural Productivity Enhancement Programme (Uganda)
APiDIA	Association des Producteurs Importateurs Distributeurs d'Intrants Agricoles (Guinea)
ARI	African Rice Initiative
ASK	Agricultural Society of Kenya
ASSEMA	Association Semencière du Mali
AVRDC	World Vegetable Center
BMS	Banque Malienne de Solidarité
BNDA	Banque Nationale de Développement Agricole (Mali)
BV Lac	Projet de Protection des Bassins Versants du Lac Alaotra (Madagascar)
CABI	CAB International
CALA	Complexe Agricole du Lac Alaotra (Madagascar)
CBO	community-based organization
CCIAK	Coopérative de Commercialisation des Intrants Agricoles de Kindia (Guinea)
CFA	currency used in 14 West and Central African countries
CFC	Common Fund for Commodities
CIAT	International Center for Tropical Agriculture
CIMMYT	International Maize and Wheat Improvement Center

CIP	International Potato Center
CMDT	Compagnie Malienne de Textile (Mali)
CMS	Centre de Multiplication de Semence (Madagascar)
CONASEM	Conseil National des Semences (Madagascar)
CTAC	Comité Technique d'Admission des Variétés au Catalogue (Madagascar)
DANIDA	Danish International Development Agency
DAS	Department of Agricultural Services (The Gambia)
DFID	Department for International Development (UK)
DNA	Direction Nationale de l'Agriculture (Guinea and Mali)
DRDR	Direction Régionale de Développement Rural (Madagascar)
EAC	East African Community
ECOWAS	Economic Community of West African States
ELISA	enzyme-linked immunosorbent assay
EPIC	établissement public à caractère industriel et commercial
EU	European Union
FAO	Food and Agriculture Organization of the United Nations
FCFA	Franc Communauté Financière Africaine
FFS	farmer field school
FICA	Farm Inputs Care Centre (Uganda)
FIFAMANOR	Centre for rural development and applied research (Madagascar)
FIOVA	Association of enterprises of the Vakinankaratra region (Madagascar)
FOFIFA	Centre National de Recherche Appliquée au Développement Rural (Madagascar)
Gamcell	Gambia Telecommunications Cellular Company Ltd
GASTA	Gambia Agrochemicals and Seed Trade Association
GDP	gross domestic product
GIC	groupement d'initiatives communes (Cameroon)
GNF	Guinean franc
GPS	Groupement de Paysans Semenciers
GRTS	Gambia Radio and Television Services (The Gambia)
GTZ	Deutsche Gesellschaft für Technische Zusammenarbeit (Germany)
HYV	high-yielding varieties
ICIPE	African Insect Science for Food and Health
ICRISAT	International Crops Research Institute for the Semi-Arid Tropics
IDEA	Investment in Developing Export Agriculture (Uganda)
IER	Institut d'Economie Rurale (Mali)
IFAD	International Fund for Agricultural Development
IFDC	International Centre for Soil Fertility and Agricultural Development
IITA	International Institute of Tropical Agriculture
IMF	International Monetary Fund
INCOFIN	Belgian investor in micro-finance
INRA	Institut National de la Recherche Agronomique (Morocco)
IPM	integrated pest management
IRAD	Institute of Agricultural Research for Development (Cameroon)
IRAG	Institut de Recherche Agronomique de Guinée (Guinea)

IRRI	International Rice Research Institute
ISTA	International Seed Testing Association
IWMF	International Women's Media Foundation
KARI	Kenya Agricultural Research Institute
KEPHIS	Kenya Plant Health Inspectorate Service
KSh	Kenyan shilling
LABOSEM	Laboratoire de Semences (Mali)
MARKETS	USAID-funded project linking farmers to markets (Nigeria)
MEADEN	Mission d'Études pour l'Aménagement et le Développement de la Province du Nord (Cameroon)
MUDIFESOF	Mumias District Federation of Soybean Farmers (Kenya)
NAADS	National Agricultural Advisory Services (Uganda)
NARI	National Agricultural Research Institute (The Gambia)
NARO	National Agricultural Research Organization (Uganda)
NASC	National Agricultural Seed Council (Nigeria)
NASECO	Nalweyo Seed Company (Uganda)
NAERLS	National Agricultural Extension and Research Liaison Service (Nigeria)
Nerica	New Rice for Africa
NGO	non-governmental organization
NSCS	National Seed Certification Service (Uganda)
OAPI	African Intellectual Property Organization
OECD	Organisation for Economic Co-operation and Development
OPSS	Opération Production Semences Sélectionnées (Mali)
OPV	open-pollinated variety
ORTM	Office de Radiodiffusion – Télévision du Mali
OSRO	FAO's Office of Special Relief Operations
PAFISEM	Projet d'appui à la filière semencière (Mali)
PBAK	Plant Breeders Association of Kenya
PSDR	Projet de Soutien au Développement Rural (Madagascar)
PVS	participatory varietal selection
QDS	quality declared seed
RIFAN	Rice Farmers Association of Nigeria
RNM	Radio National Malagasy
SACCO	saving and credit cooperative (Uganda)
SCAA	Société Commerciale et Agricole d'Ambohimangakely (Madagascar)
SCSP	Service de Contrôle de Semences et Plants (Morocco)
SEEDAN	Seed Association of Nigeria
SEMRY	Société d'Expansion et de Modernisation de la Riziculture de Yagoua (Cameroon)
SG2000	Sasakawa Global 2000
SMV	Service des Semences et Matériel Végétal (Madagascar)
SNPRV	Service National de la Promotion Rurale et de la Vulgarisation (Guinea)
SOC	Service Officiel de Contrôle des Semences (Madagascar)
SOGETA	Société de Gestion des Terres Agricoles (Morocco)
SONACOS	Société Nationale de Commercialisation de Semences (Morocco)

SRI	system of rice intensification
STAK	Seed Trade Association of Kenya
TLS	truthfully labelled seed
TRIAS	Belgian NGO (merger between ACT, Ieder voor Allen and Form)
TV	television
UCSCU	Uganda Cooperative Savings and Credit Union
UNADA	Uganda National Agro-input Dealers Association
UNDP	United Nations Development Programme
UPOV	Union for the Protection of New Varieties of Plants
USAID	US Agency for International Development
USDA	US Department of Agriculture
USTA	Uganda Seed Trade Association
VAT	value added tax
VKMMV	Union of seed producer cooperatives in Vakinankaratra (Madagascar)
VSO	Voluntary Service Overseas
WASA	West Africa Seed Alliance
WFP	World Food Programme
ZARDI	Zonal Agricutural Research and Development Institute (Uganda)

Glossary of Terms

breeder seed	Seed that is produced by plant breeders in small quantities for multiplication to reach the desired volumes for sale to companies or outgrowers
certified seed	Seed that has been verified to be varietally pure, clean and viable (high germination rate)
cross-pollinating species	A plant species that exchanges genetic material in the form of pollen from one plant to another (as opposed to self-pollinating species). Examples of cross-pollinated crops are maize, pearl millet, sunflower, lucerne and forage grasses
fonio	One of the world's fastest growing cereals, reaching maturity in as little as 6 to 8 weeks. It is a nutritious crop that grows in semi-arid areas with poor soils, where rains are brief and unreliable. The grains are used in porridge and couscous, for bread and for beer
formal seed sector	The regulated chain of seed production and marketing involving scientific plant breeding, multiplication following established procedures, processing, bagging, labelling and marketing
foundation seed	Seed produced by a research centre, a company or outgrowers that is one generation after breeder seed but will be multiplied again before being sold to farmers
gombo	Okra. A traditional food plant in Africa, from the hibiscus family, that is valued for its edible green fruits. When the seed pods are cooked, they result in a characteristic 'goo' or slime
hybrid seed	Seed produced by crossing two or more separate inbred lines. Hybrid seed typically produces high yields the first year, but the yield drops if recycled for a second year. The percentage yield loss that occurs in recycling depends on the type of hybrid and the growing environment
informal seed sector	The chain of seed production and marketing involving farmers who save seed from harvest to planting, occasionally selling or exchanging seed with other farmers, but without any official testing
open-pollinated variety	A crop variety that can pollinate and reproduce naturally (through either self- or cross-pollination) and whose seed can be saved and reused over various seasons, as opposed to hybrid seed

seed system	The set of interconnected institutions involved in developing, producing, testing and marketing seed. It involves both the formal and informal sector
self-pollinating species	A plant species in which the majority of pollination occurs within the same plant (as opposed to cross-pollinating species). Examples of self-pollinating species are wheat, rice, chickpeas and groundnuts. Sorghum, cowpea, beans and pigeon pea are predominantly self-pollinating
standard seed	A category of seed in some seed quality control schemes which may be tested but is not subject to full certification procedures
truthfully labelled seed	A category of seed that is not certified, but is labelled according to the characteristics and origin of the seed
varietal release	A procedure by which a committee reviews the results of field trials and decides whether a new variety can be named and made available for sale to farmers
vegetatively propagated species	A species that does not produce viable seed, but instead reproduction occurs by using part of an existing plant. Examples of vegetatively propagated species include cassava, sweet potato, potato and bananas. Cassava and sweet potato are propagated by using portions of the stem, potato by using tubers and banana by cutting and replanting 'suckers' (small shoots) from the banana plant
ware potatoes	Potatoes to eat
yam	Perennial herbaceous vine that yields large, edible starchy tubers

Index